KNOWLEDGE

AND

POWER

KNOWLEDGE

AND

POWER

....

THE INFORMATION THEORY
OF CAPITALISM AND HOW IT IS
REVOLUTIONIZING OUR WORLD

....

GEORGE GILDER

Since 1947
REGNERY
Publishing, Inc.
An Eagle Publishing Company • Washington, DC

Cataloging-in-Publication data on file with the Library of Congress

ISBN 978-1-62157-027-1

Published in the United States by
Regnery Publishing, Inc.
One Massachusetts Avenue NW
Washington, DC 20001
www.Regnery.com

Manufactured in the United States of America

10 9 8 7 6 5 4 3 2 1

Books are available in quantity for promotional or premium use. Write to Director of Special Sales, Regnery Publishing, Inc., One Massachusetts Avenue NW, Washington, DC 20001, for information on discounts and terms, or call (202) 216-0600.

Distributed to the trade by
Perseus Distribution
250 West 57th Street
New York, NY 10107

For David Rockefeller, Ph.D. in economics,
Hayek tutee, who taught me the limits of knowledge and power
and the valor of virtue

Contents

"While market economies are often thought of as money economies, they are still more so knowledge economies.... Economic transactions are purchases and sales of knowledge."

"After all, the cavemen had the same natural resources at their disposal as we have today.... We are all in the business of buying and selling knowledge from one another, because we are each so profoundly ignorant of what it takes to complete the whole process of which we are a part."

"'How could we have gone so wrong?' ... The short answer is that power trumps knowledge."

—THOMAS SOWELL, *Knowledge and Decisions*, 1979 (P. 47),
AND *Basic Economics*, 2007 (P. 424)

• • • •

"Life is plastic, creative! How can we build this out of static, eternal, perfect mathematics? We shall use post-modern math, the mathematics that comes after Gödel, 1931, and Turing, 1936, open not closed math, the math of creativity...."

—GREGORY CHAITIN, *Proving Darwin*, 2012 (P. 14)

Foreword

A Venture Investor from Bell Labs Channels the Noise and the Knowledge

I GET OFF THE GREEN Number 5 train at the Wall Street stop as I've done a million times before. It could be this year or years ago. It doesn't matter. I carefully shuffle out with the pack of humanity, half in suits, the other half wearing bike messenger bags; make my way out through the turnstiles; and, shoulder to shoulder with people in a hurry, slip and slide up the litter-strewn stairs onto the Street.

The sun is bright, piercing. You can see everything, but none of it comes into focus. I'm instead distracted by the racket. Cars honking, jack hammers rattling away, a guy selling the *New York Post* going on about the latest sensational Twittered sex crime. Trucks roar by me. Subways screech beneath me. An ambulance with sirens blaring goes up on the sidewalk to get around some construction. Bad music—someone rapping "Turn it up! Bring the Noise"—is blaring from a Starbucks that turns out to have no restroom. My head is spinning. I can hardly hear myself think.

But underneath all that noise, I hear a sound, sort of a *thub-dub, thub-dub*, a relentless reggae beat, sometimes loud, sometimes soft, faster, slower, but the pulse is always there. Is it a heartbeat, the predictable rhythm of life? Or does it bear a signal, a difference, a delta of news? The precious modulation of a wave of new creativity in the channels of the economy?

I've got "five hundred large" of other people's money to invest. "I won't lose any of your capital and I'll find the next Microsoft," I told them in 1995. What the hell was I thinking? It's so loud down here it's hard to make sense of anything. Every story sounds good and every stock looks like a bargain—but there are so many stories, they drown each other out. Too many stories essentially merge into one endless market oscillation—a random motion through time that will deceive most technical analysts who take it for a signal and will be left gasping and grasping for handfuls of noise.

I think I'm different. I've got alpha, baby—which in Wall Street–speak means I think I can generate excess returns over the market. I think I can find the profits of surprise, the yield of real knowledge. Everyone says that, of course, but most investors are all beta—just volatility, just the random motion of the surf. When markets go up the beta warriors outperform, and when markets go down they get killed.

To generate alpha, I need help, direction, signposts, analysts, and sometimes even brandy-toting salesmen. But about the only pointer in view is George Washington's outstretched arm at Wall and Broad, aiming across the street to the New York Stock Exchange—almost as a warning to watch out for those guys in funny-colored blazers. On the other corner is 23 Wall Street, the J. P. Morgan headquarters bombed by anarchists in 1920. Now they blow these banks up from the inside—with combustible illusions of alpha.

I've got to put that money to work, buy stocks that go up five to ten times, and prove that my alpha is real. In this book, and in Claude Shannon's classic model that it describes, alpha goes under the name of "entropy." But it's essentially the same thing. It is the unanticipated signal, the upside surprise, the unexpected return, the messages among the noise on the Street. The predictable returns are already in prices, in interest rates. I have to achieve upside surprises that are not implicit in current prices and I have to get them not merely today or tomorrow, but month by month, year by year. And I have to hedge them with shorts of stocks that are overblown and going down faster than the *Titanic*. Simple enough, right? I wish.

Back on the subway this morning, I was channeling Larry, a guy in a leisure suit who ran "go-go money," as we used to call it, back in 1973. "Ah," he tells me, "those were the days and daze. A White Weld institutional salesman would call me every morning at nine with the early word on what his analysts were saying on Polaroid or Xerox or Philip Morris. Trading cost seventy-five cents a share, but who cares, there were only fifty stocks that mattered, the Nifty Fifty, and you just bought 'em, never sold. Maybe I'd get some ideas from the 'Heard on the Street' in the *Journal*, or maybe 'Inside Wall Street' from *Business Week*."

Unfortunately, the Nifty Fifty melted into a worthless heap, and Vanguard, John Bogle's pioneering new fund, rose from the ashes. Propelled by a Big Bang of market deregulation, negotiated commissions, and lower transaction costs, Vanguard back in 1975 figured that alpha was a myth, that no mere mortals could beat the market, so they indexed the whole damn thing. Buying a Vanguard fund, you merely bought a statistical sample of the market. It was like driving all the knowledge out of prices. Danny Noonan in *Caddy Shack* was told to "be the ball;" Vanguard told us to "be the market." But if we are the market, we do not shape it; we are just bounced and dribbled around. Shannon, the ultimate alpha man of investing, as we learn in this book, would not have been amused.

Twenty-five years later, much of the market is mindlessly indexed. That means it is all beta. The knowledge is leaching away in the surf of noise and rapid trading. Computers in, humans out; this is classic 1970s sci-fi made all too real. A scream from a homeless man playing Angry Birds on his iPhone ends my subway séance with the wisdom of the 1970s.

An index is the market. It's a carrier, a channel, as defined mathematically by Shannon at Bell Labs in his seminal work on information theory. An index can yield only the predictable market return, mostly devoid of the profits of creativity and innovation, which largely come from new companies outside the index. I had to beat the indexes—by a lot. That means I needed knowledge. Riding on the channel, knowledge portends deformation of the mean. It is signaled by surprise, upside and downside, but it is not realized until the surprise—the information—is understood.

As the information revolution described in this book began to take off, I had an advantage. I started my career at Bell Labs, thirty-five years after Shannon. On your first day there, you are issued a nine-by-twelve brown leatherette bag with a Bell logo in the lower corner. There were guards at every entrance and exit making sure employees didn't, uh, liberate equipment from the Labs. But the rule was that the guards would not search your Bell Bag.

In the days before personal computers, Bell Labs employees—OK, by that I mean me!—tried to take home a Digital Equipment PDP-11 mini-computer by taking it apart and fitting it into their Bell Bag, much as *M*A*S*H*'s Radar O'Reilly shipped home a Jeep. Rumor has it that the Bag was the reason Shockley and others invented the transistor. Machines made out of vacuum tubes didn't fit—too much material, not enough information. At Bell Labs we were reducing everything to information. Today it is almost all information, and you could steal its crown jewels of software in a thumb drive.

Anyway, in a few years I left Bell Labs and moved to Wall Street.

As I strolled down Wall Street, the *thub-dub* was getting louder. It was the market, the pulse of the street. It's what everyone thinks. Every day, you're hit with a fire-hose blast of information—in the *Wall Street Journal*, on Yahoo! Finance, in real-time stock quotations, in press releases, on StockTwits.

But I still don't have knowledge, interpreting the surprises that others don't know about, that will drive a new narrative. You have to work and think and stress and fret to surmise the surprises by first fathoming the pulse.

The battle is just filtering out the few tiny gems, the insights that make up the new knowledge. If not, my "five hundred large" gets returned to the index cesspool. *Thub-dub* this.

Except in a few exceptional periods of a bubble market, if there is no noise, there is no return. If it's so painfully obvious, like the Nifty Fifty of the '70s; if retired couples are talking about buying more Apple shares in the quiet of an airport Admiral's Club; run away until the noise returns.

As an investor, I need to feel the pulse every day and wade through the drivel in order to pan the gold. The pulse has to reverberate in my veins, but only so I understand what the market is saying *today*. Then I have to resist the calming effect of that *thub-dub* of conventional thinking and venture out into the noise, out on the edge, to find new information and what's next, which can lead to knowledge. It's as elusive as humpback whales, but it's there.

Amid the clutter of trends running around the Street, though, it is hard to tell what is real and what is just Synsonic synthesized sound. "Reg FD"—regulation full disclosure—means companies only give "guidance" on how they see business tracking once a quarter, on an earnings release conference call with questions like "Congratulations on the great quarter, uh, what's your tax rate going forward?" That is what Shannon might call zero-entropy communication. It removes information from the market when I need more and more.

With a beta of 1.0, any sample of the market exactly recapitulates the market averages. It's the insight extracted from the information—that alpha—that separates the winners from the snoozers on Wall Street. Indexing is a waste heap—information so merged and muffled that it hides knowledge rather than reveals it. All beta, no alpha.

So what do the best modern money managers do? They live for the pulse, in the pulse, but then they work out, often by an educated gut instinct, what is different and what is going to change—*where the surprises will come*—where Shannon hid the entropy. That's valuable knowledge. No more leisure-suit Nifty Fifty, no more indexing, no more day traders, no more momentum investing, and no more fooling around.

So I went to Palo Alto in the midst of Silicon Valley. Why Silicon Valley, where a 1,500-square-foot house runs $2.5 million? Because it's where the surprises are. Beating the market turned out to have nothing to do with trading or the plumbing of Wall Street. It had to do with understanding and predicting the surprises, the changes, and the productivity fabric of the economy. The rest is noise.

Every day in Silicon Valley, someone writes a clever piece of code that changes retail or uses information theory to write a security algorithm or invents a new way to shape Wi-Fi beams. These are all surprises. Getting away from the scopes of stock market trading and into the microscopic detail of how technology is changing and its effect on human-machine interfaces and why many existing industries will collapse is the only way to gain real actionable knowledge.

A day doesn't pass that I'm not surprised. Many years ago I met with a team that could very cheaply jam five gigabits of information per second down a couple of meters of cable, unheard of at the time. I didn't know you could do that, but *voilà*, HDMI (high-definition multimedia interface) was born. I can almost guarantee it is how you get high-def video to your flat screen TV. A game changer—not overnight, but over years. It might not have been the next Microsoft, but it was good enough. It went from noise to the narrative, the pulse, and huge amounts of wealth were created.

No index can capture that. The index is retrospective. The crucial alpha, the entropy, the signal modulating that linear advance—information light enough to stash away in your Bell Bag or thumb drive and shape the future—comes from knowledge of the entrepreneurial surprises harbored on the edge of the noise.

The big narrative of the economy changes daily. That's productivity and progress. It is that high-energy message that the market as medium carries into the future.

—*Andy Kessler*

PART ONE

The Theory

The Need for a New Economics

MOST HUMAN BEINGS understand that their economic life is full of surprises. We cannot predict the value of our homes or prices on the stock market from day to day. We cannot anticipate illness or automobile accidents, the behavior of our children or the incomes of our parents. We cannot know the weather beyond a week or so. We cannot predict what course of college study will yield the best lifetime earnings or career. We are constantly startled by the news. We are almost entirely incapable of predicting the future.

Yet economics purports to be strangely exempt from this fact of life. From Adam Smith's day to our own, the chief concern of the discipline has been to render economic events unsurprising. Given a supply x of corn and a demand y, the price will be z. Change x or y and hold all else equal and the price will instead be a predictable z. The discernment of orderly rules governing the apparent chaos of life was a remarkable achievement and continues to amaze. Economists such as Steven Leavitt of *Freakonomics* fame and Gary Becker of the University of Chicago became media stars for their uncanny ability to unveil what "we should have known."[1] Closer investigation, however, reveals that even these ingenious analysts are gifted chiefly with 20/20 hindsight. They prosper more by explaining to us what has happened than by anticipating the future with prescient investments.

The passion for finding the system in experience, replacing surprise with order, is a persistent part of human nature. In the late eighteenth century, when Smith wrote *The Wealth of Nations*, the passion for order found its fulfillment in the most astonishing intellectual achievement of the seventeenth century: the invention of the calculus. Powered by the calculus, the new physics of Isaac Newton and his followers wrought mathematical order from what was previously a muddle of alchemy and astronomy, projection and prayer. The new physics depicted a universe governed by tersely stated rules that could yield exquisitely accurate predictions. Science came to mean the elimination of surprise. It outlawed miracles, because miracles are above all unexpected.

The elimination of surprise in some fields is the condition for creativity in others. If the compass fails to track North, no one can discover America. The world shrinks to a mystery of weather and waves. The breakthroughs of determinism in physics provided a reliable compass for three centuries of human progress.

Inspired by Newton's vision of the universe as "a great machine," Smith sought to find similarly mechanical predictability in economics. In this case, the "invisible hand" of market incentives plays the role of gravity in classical physics. Codified over the subsequent 150 years and capped with Alfred Marshall's *Principles of Economics*, the classical model remains a triumph of the human mind, an arrestingly clear and useful description of economic systems and the core principles that allow them to thrive.

Ignored in all this luminous achievement, however, was the one unbridgeable gap between physics and any such science of human behavior: the surprises that arise from free will and human creativity. The miracles forbidden in deterministic physics are not only routine in economics; they constitute the most important economic events. For a miracle is simply an innovation, a sudden and bountiful addition of information to the system. Newtonian physics does not admit of new information of this kind—describe a system and you are done. Describe an economic system and you have described only the circumstances— favorable or unfavorable—for future innovation.

In Newton's physics, the equations encompass and describe change, but there is no need to describe the agent of this change, the creator of new information. (Newton was a devout Christian but his system relieved God or his angels of the need to steer the spheres.) In an economy, however, everything useful or interesting depends on agents of change called entrepreneurs. An economics of systems only—an economics of markets but not of men—is fatally flawed.

Flawed from its foundation, economics as a whole has failed to improve much with time. As it both ossified into an academic establishment and mutated into mathematics, the Newtonian scheme became an illusion of determinism in a tempestuous world of human actions. Economists became preoccupied with mechanical models of markets and uninterested in the willful people who inhabit them.

Some economists become obsessed with market efficiency and others with market failure. Generally held to be members of opposite schools— "freshwater" and "saltwater," Chicago and Cambridge, liberal and conservative, Austrian and Keynesian[2]—both sides share an essential economic vision. They see their discipline as successful insofar as it eliminates surprise—insofar, that is, as the inexorable workings of the machine override the initiatives of the human actors.

"Free market" economists believe in the triumph of the system and want to let it alone to find its equilibrium, the stasis of optimum allocation of resources. Socialists see the failures of the system and want to impose equilibrium from above. Neither spends much time thinking about the miracles that repeatedly save us from the equilibrium of starvation and death.

The late financial crisis was perhaps the first in history that economists actually caused. Entranced by statistical models, they ignored the larger dimensions of human creativity and freedom. To cite an obvious example, "structured finance"—the conglomerations of thousands of dubious mortgages diced and sliced and recombined and all trebly insured against failure—was supposed to eliminate the surprise of mortgage defaults. The mortgage defaults that came anyway and triggered the collapse came not from the aggregate inability of debtors to pay as the economists calculated, but from the free acts of homebuyers. Having bet on constantly rising home prices, they simply folded their hands and walked away when the value of their houses collapsed. The bankers had accounted for everything but free will.

The real error, however, was a divorce between the people who understood the situation on the ground and the people who made the decisions. John Allison is the former CEO of a North Carolina bank, BB&T, which profitably surmounted the crisis after growing from $4.5 billion in assets when he took over in 1989 to $152 billion in 2008. Allison ascribed his success to decentralization of power in the branches of his bank.

But decentralized power, he warned, has to be guarded from the well-meaning elites "who like to run their system and hate deviations." So as CEO, Allison had to insist to his managers that with localized decision-making, "We get better information, we get faster decisions, we understand the market better."[3]

Allison was espousing a central insight of the new economics of information. At the heart of capitalism is the unification of knowledge and power. As Friedrich Hayek, the leader of the Austrian school of economics, put it, "To assume all the knowledge to be given to a single mind…is to disregard everything that is important and significant in the real world."[4] Because knowledge is dispersed, power must be as well. Leading classical thinkers such as Thomas Sowell and supply-siders such as Robert Mundell refined the theory.[5] They all saw that the crucial knowledge in economies originated in individual human minds and thus was intrinsically centrifugal, dispersed and distributed.

Enforced by genetics, sexual reproduction, perspective, and experience, the most manifest characteristic of human beings is their diversity. The freer an economy is, the more this human diversity of knowledge will be manifested. By contrast, political power originates in top-down processes—governments, monopolies, regulators, and elite institutions—all attempting to quell human diversity and impose order. Thus power always seeks centralization.

The war between the centrifuge of knowledge and the centripetal pull of power remains the prime conflict in all economies. Reconciling the two impulses is a new economics, an economics that puts free will and the innovating entrepreneur not on the periphery but at the center of the system. It is an economics of surprise that distributes power as it extends knowledge. It is an economics of disequilibrium and disruption that tests its inventions in the crucible of a competitive marketplace. It is an economics that accords with the constantly surprising fluctuations of our lives.

In a sense, I introduced such an economics more than thirty years ago in *Wealth and Poverty* and reintroduced it in 2012 in a new edition. That book spoke of economics as "a largely spontaneous and mostly unpredictable flow of increasing diversity and differentiation and new products and modes of production ... full of the mystery of all living and growing things (like ideas and businesses)." Heralding what was called "supply-side economics" (for its disparagement of mere monetary demand), it celebrated the surprises of entrepreneurial creativity. Published in fifteen languages, the original work was read all around the globe and reigned for six months as the number one book in France. President Ronald Reagan made me his most-quoted living author.

In the decades between the publications of the two editions of *Wealth and Poverty*, I became a venture capitalist and deeply engaged myself in studying the dynamics of computer and networking technologies and the theories of information behind them. In the process, I began to see a new way of addressing the issues of economics and surprise.

Explicitly focusing on knowledge and power allows us to transcend rancorous charges of socialism and fascism, greed and graft, "voodoo economics" and "trickle-down" theory, callous austerity and wanton prodigality, conservative dogmatism and libertarian license.

We begin with the proposition that capitalism is not chiefly an incentive system but an information system. We continue with the recognition, explained by the most powerful science of the epoch, that information itself is best defined as surprise—what we cannot predict rather than what we can. The key to economic growth is not acquisition of things by the pursuit of monetary rewards but the expansion of wealth through learning and discovery. The economy grows not by manipulating greed and fear through bribes and punishments but by accumulating surprising knowledge through the conduct of the falsifiable experiments of free enterprises. Crucial to this learning process is the possibility of failure and bankruptcy.

Because the system is based more on ideas than on incentives, it is not a process that is changeable only over generations of Sisyphean effort. An economy is a "noosphere" (a mind-based system), and it can revive as quickly as minds and policies can change.

That new economics—the information theory of capitalism—is already at work in disguise. Concealed behind an elaborate mathematical apparatus, sequestered by its creators in what is called information technology, the new theory drives the most powerful machines and networks of the era. Information theory treats human creations or communications as transmissions through a channel, whether a wire or the world, in the face of the power of noise, and gauges the outcomes by their news or surprise, defined as "entropy" and consummated as knowledge. Now it is ready to come out into the open and to transform economics as it has already transformed the world economy itself.

The Signal
in the Noise

I FIRST ENCOUNTERED the information theory at the center of the contemporary economy of capitalism in 1993 during a trip into the sandy hills of La Jolla, California, north of San Diego.

I came to visit Qualcomm Corporation, a company founded eight years before. By computerizing the communications of all the mobile devices you use every day—your cell phone, iPad, Kindle, or netbook—Qualcomm has become one of the world's most valuable and influential corporations. It reached a market capitalization of over $110 billion in 2012, surpassing Intel as the most highly valued U.S. microchip producer. But in the early 1990s, it aroused the kind of enmity usually reserved for tobacco companies.

Writing articles every month for the new technology magazine *Forbes ASAP*, I found myself surrounded by ardent enemies of this apparently innocent wireless vendor. Highly placed executives and consultants—and even the occasional engineer or scientist—urged me to expose the conspiracy of a fanatical cult led by Qualcomm to fool the world into

adopting what they called its impossibly complex and physically imprac-
tical digital wireless technology. While I was giving a speech in Germany,
a fervent Qualcomm opponent actually interrupted me from the floor,
warning my audience of European telecom executives against my sedi-
tious message that Qualcomm's technology would prevail.

The usual charge against Qualcomm's system was that it "violates
the laws of physics." So Bruce Lusignan, a learned professor of electrical
engineering at Stanford, informed me. A man with sixteen patents in
signal processing and related fields, Lusignan generally knows what he
is talking about. The laws of physics, he pointed out, "actually favor
analog transmission over digital." If as much investment had been made
in improving the existing system as was lavished on digital, he said, the
future of cell phones would be analog.

Lusignan was right about the laws of physics. Analog signals reproduce the full sound waves of voices in the form of full electrical waves
rather than waves sampled twice a cycle or hertz for a numerical approx-
imation of the sound. The analog transmission is radically more efficient
for transmitting sounds, and at the time, it accounted for more than 60
percent of all U.S. cell phone service.

What Lusignan missed was the effect of the laws of information, with
which Qualcomm had overcome the physical laws. In our time, the intellectual prestige of physics, at least among non-scientists, is supreme. But
for conveying *information*, physical models are relatively impoverished
compared with chemical models, which in turn compare poorly with the
biological. A few thousand lines of genetic code (a tiny fraction of any
organism's genome) convey more information than anything in the realm
of physics.

We admire physics because, compared with biology, it is relatively
complete. We know pretty well how the solar system works; the immune
system baffles us. We split the atom before we cured polio. Physics is
more complete precisely because the information content of the system
is so limited. Killing a virus without killing the man who carries it turns
out to be a vastly more complex and information-intensive exercise than
orbiting the planet, exploring Mars, or incinerating Hiroshima. The latter task, recall, needed only an airplane driven by a propeller and an

internal combustion engine and a bomb constructed in less than five years to be accomplished.

Physics is not the final word. Qualcomm triumphed by moving beyond physics to the new science of information, transforming the physical scarcity of "bandwidth" into an abundance of wireless communications.

"Bandwidth" is the apparent physical carrying capacity of a connection, whether wire, air, cable, fiber optic web of light, or dark telecom "cloud." At the receiving end, we must be able to distinguish between the signal and the "noise"—the word and the wire. If content is to get through, the payload must be separable from its packaging.

In biology, Francis Crick dubbed this proposition the Central Dogma: information can flow from the genetic message to its embodiment in proteins—from word to flesh—but not in the other direction. Similarly, in communications, any contrary flow of influence, from the physical carrier to the content of the message, is termed noise.

One way to enhance transmission is by eliminating noise: making the channel as stable as possible so that every modulation of the carrier can be interpreted as "signal." We communicate through the physical contrast between silent channel and loud signal. Qualcomm would change all this, seeking not to eliminate noise but to transcend and transform it into information. Mastery of the permutations of noise, as I was to discover, is central to the achievements of Qualcomm and the insights of information theory.

Before my trip to Qualcomm, my chief enthusiasm in technology was the physics of silicon. In 1989 I had written a book called *Microcosm: The Quantum Era in Science and Technology*, which used physics to understand the dynamics of the new semiconductor industry. I liked to cite Blake's poetic vision of seeing "worlds in grains of sand," which I took to anticipate the microchip, inscribing vast webs of intricate circuitry on slivers of opaque silicon. I extended the vision into "spinning out the grains of sand around the world" in worldwide webs of glass and light. I believed that the transparent silicon of fiber optics was opening a new and unprecedented promise of bandwidth abundance. In both cases, mastery of the physical characteristics and behavior of silicon, making it predictable and controllable, laid the foundation for an industry in which creativity constantly surprises.

With the encouragement of a fiber optics pioneer named Will Hicks and an IBM engineer named Paul Green, I suggested in 1991 that these worldwide webs of glass and light—with bandwidths millions of times greater than those possible with copper wires—would usher in a new era of economics. Fiber optics enabled an all but limitless broadband flow of information between peoples once linked chiefly by narrow seaborne channels of trade and noisy copper cables. Webs of glass would achieve a new economics of abundance. I dubbed this the "fibersphere" and I conceived it as primarily an achievement of quantum physics and its engineering derivative, solid-state chemistry.

I soon realized, however, that to serve mobile human beings wherever they moved, the fibersphere would need the atmosphere as your lungs need air. And in the atmosphere, bandwidth was far less abundant. It would not be possible to compete with the sun in San Diego in transmitting photonic signals through the air. Restricted to frequencies outside the hyper-broadband blast of sunlight, bandwidth in the atmosphere would face daunting limits. This scarcity of bandwidth was the catalyst for information theory, which became the foundation for wireless communications.

At the time people were warning me about Qualcomm, I knew little about the company or about information theory. But I thought I should visit Qualcomm's headquarters before some physics professor in Palo Alto put its executives under citizen's arrest.

At a small table overlooking the atrium of Qualcomm's new headquarters, the company's founders, Andrew Viterbi and Irwin Jacobs, tried to explain their controversial technology to me. Jacobs was tall, lanky, and soft-spoken. He used homely analogies to describe the virtues of Qualcomm's solution. Viterbi was short and paunchy and determined to expound the decisive points from information theory. There was a discernible tension between the two, one talking down to me and one talking up. I was not surprised when Viterbi left the company less than a decade later.

Both men possessed intellects far superior to those of most executives I met, even in cerebral Silicon Valley. Jacobs was clearer in explaining his system to me and was more quotable for my articles in *Forbes*. As he later explained to me, he went to MIT in the mid-1950s to study the physics and engineering of electromagnetism. But all the excitement at the time surrounded Claude Shannon, the "playful polymath" (in the words of John Horgan) who had first identified the laws of information theory less than a decade earlier. Jacobs ended up studying information theory with Paul Elias, Robert Fano, and Shannon, and, when Jacobs became a professor, his office was just down the corridor from Shannon's.

It was Viterbi, however, who posed for me a profound riddle of information theory that launched me on a twenty-year exploration of Shannon's ideas, from communications to biology and on to economics. Viterbi earnestly identified the secret of Qualcomm's superiority as the recognition that a communications system is most capacious and efficient when its contents most closely resemble not a clear channel and decisive signal but a fuzzy stream of "white noise."

What could he have meant? I stubbed my neurons on the idea of noise as a carrier of information. Viterbi's view seemed to wrap the Qualcomm riddle in a mystery. Surely noise is the opposite of communication, and "white" means that the racket is equally dispersed among all frequencies, or "colors," of noise, enveloping the mystery in an enigma of uniform static—to complete the Churchillian image.

Viterbi's statement not only defied common sense, but it also contradicted what nearly everyone else I talked to in the industry said. That might explain why the rest of the industry was so resistant to Qualcomm. All telecom was engaged in a war against noise, laboring to banish it, suppress static, enhance signal-to-noise ratios, and jack up the volume of the signal to overcome the buzz. The industry was coalescing around digital transmission standards that broke up the signal stream into time slots and assigned each slot to one message alone with no noise from other transmissions.

The favored digital system was time division multiple access (TDMA), which was popular in Europe. Telephone companies liked TDMA because they already used it to share or multiplex all their wire-line links, which did not have to deal with the vagaries of mobile communications. By encapsulating each packet of data in an exclusive slot of time and frequency, TDMA shielded its packets from interference.

This virtue, however, made TDMA a relatively rigid and inefficient system because it wasted all its unused time slots. (Most access phone wires are empty most of the time, after all). TDMA allows precious time slots to pass irretrievably by like empty freight cars receding down the tracks. Moreover, because the traditional strategy was to shout across an exclusive channel (in the case of TDMA, only momentarily exclusive), the "walls" of TDMA's slots had to be thick. This meant taking more bandwidth for insulation, so that next-door neighbors' domestic incidents did not come through as noise.

Qualcomm's system—called code division multiple access, or CDMA—was completely different. Rather than speaking more loudly to make themselves clear over longer distances, all communicators would speak more quietly. Jacobs likened Qualcomm's scheme to a cocktail party in which each pair of communicators spoke its own language. They would differentiate their calls not through time slots or narrow frequency bands but through codes. Spread across the available spectrum, these codes would resemble white noise to anyone without a decoder.

In *An Introduction to Information Theory: Symbols, Signals, and Noise*, John R. Pierce, Shannon's close colleague at Bell Labs (and coiner of the word "transistor"), puts numbers on this multilingual cocktail party.[1] Engineers have a choice between two strategies to maximize channel capacity: they can increase the bandwidth of the signal or they can increase its power-to-noise ratio. Most of the industry was seeking to enhance the signal-to-noise ratio, speaking more loudly to be heard more clearly. As James Gleick commented in his definitive history, *The Information*, "Every engineer, when asked to push more information through a channel, knew what to do: *boost the power*."[2] But as Pierce showed in 1980, doubling the bandwidth of the signal from four megahertz to eight megahertz allows for a more than thirty-three-fold drop in the power-to-noise ratio.[3] Reducing the power and expanding the bandwidth was over sixteen times more efficient in the example than increasing the signal power at the same bandwidth. There was a paradox of loudness. One person could be heard better by speaking at higher volume, but if everyone did it, communication would drown in the ambient noise.

Pierce concluded, "If we wish to approach Shannon's limit for a chosen bandwidth we must use as elements of the code long, complicated signal waves that resemble gaussian noise."[4] This was Viterbi's insight behind the success of spread-spectrum CDMA.

Shrouded in an extended code that seemed like low-level background noise, the spread-spectrum message could get through while only slightly interfering with other messages. Noise would build up incrementally in the cell as more users made calls. Allowing all the calls to use all the frequencies in the cell all the time without the wasteful rigidity of TDMA time slots or the noisy chaos of high-power analog transmissions, CDMA maximized capacity. To accommodate more users during traffic jams on the freeway, coded calls could even move into less crowded neighboring cells, because all calls and cells used the same frequencies.

With a future of wireless Internet on my mind, the Qualcomm advance excited me. I could see that CDMA would be far superior for bursty data communications that might overflow time slots or narrow frequency bands. The CDMA cocktail party would maximize communication if everyone spoke as quietly as possible in his chosen language, or code. To everyone else in the cell, the conversation would be indistinguishable from background noise. These "quasi-noise" codes would be readily translated by ever more powerful microchips in the cell phones that Qualcomm would supply or license for a reasonable fee. It was evident that, other things being equal, the Qualcomm strategy would prevail.

But the real news was better than that. For other things were not equal. In a Moore's Law world, with the cost of computing capacity falling by half every two years, the economics of silicon favored technologies that wasted computer power but conserved "physical" resources such as wireless spectrum. This was the true meaning of an "information economy" that Peter Drucker and others had merely glimpsed.

In an information economy, entrepreneurs master the science of information in order to overcome the laws of the purely physical sciences. They can succeed because of the surprising power of the laws of information, which are conducive to human creativity. The central concept of information theory is a measure of freedom of choice. The principle of matter, on the other hand, is not liberty but limitation—it has weight and occupies space.

The power of any science is limited by its information potential. We tend to overestimate the technological usefulness of physics and its laws precisely because its limited information potential makes it seem so rational and predictable. But as Shannon showed, predictability and information operate in opposition.

Transcending the laws of physics by the laws of information is not a pie-in-the-sky idea. It has been happening in the most dynamic sectors of the economy since the last century. This same step-by-step transcendence of the physical by the informational, of matter by idea, has powered all economic development through all of history and before. In previous eras, this process was less obvious because the power of information was manifested in harnessing the laws of physics rather than in transcending them. From the wheel to the Roman arch to the fragile wings lifting off the sand at Kitty Hawk, man used physics to ease the burdens of matter, to control, guide, or support more with less, to enliven matter with mind. In the information age, it is physics itself that is subdued. In the early 1990s, Qualcomm was the center of that effort.

Jacobs's cocktail-party analogy quelled my objections for the moment. But at the time I still didn't comprehend what Viterbi was telling me. I returned to the *Forbes ASAP* offices determined to plow through Shannon's information theory papers to get to the bottom of this apparent enigma of "white" or random noise and maximum information transmittal. That pursuit made me an enthusiastic supporter of Qualcomm, the best major-market American stock of the 1990s, rising in value twenty-five-fold in ten years. It also impelled me toward an information theory of capitalism that is as much a departure from standard economics as CDMA was from the prevailing protocols in the phone industry. From the equilibrium and spontaneous order of Adam Smith and his heirs, from invisible-handed markets and perfect competition, supply and demand, and rewards and punishments, I was pushed to theories of disequilibrium and disorder, and information and noise, as the keys to understanding economic progress.

The Science of Information

THE CURRENT CRISIS of economic policy cannot be understood as simply the failure of either conservative or socialist economics to triumph over its rival. It cannot be understood, as Paul Krugman or Ron Paul might wish, as a revival of the debate between the Keynesian and Austrian schools—John Maynard Keynes and Paul Samuelson against Friedrich Hayek and Ludwig von Mises. The hard science that is the key to the current crisis had not been developed when Keynes and Hayek were doing their seminal work.

That new science is the science of information. In its full flower, information theory is densely complex and mathematical. But its implications for economics can be expressed in a number of simple and intelligible propositions. All information is surprise; only surprise qualifies as information. This is the fundamental axiom of information theory. Information is the change between what we knew before the transmission and what we know after it.

From Adam Smith's day to ours, economics has focused on the nature of economic order. Much of the work of classical and neo-classical economists was devoted to observing the mechanisms by which markets, confronted with change—especially change in prices—restored a new order, a new equilibrium. Smith and his successors followed in the footsteps of Newton and Leibniz, constructing a science of systems.

What they lacked was a science of disorder and randomness, a mathematics of innovation, a rigorous measure and mandate for freedom of choice. For economics, the relevant science has arrived just in time. The great economic crisis of our day, a crisis of theory as well as practice, is a crisis of information. It can be grasped and resolved only by an economics of information. Pioneered by such titans as Kurt Gödel, John von Neumann, and Alan Turing, the mathematical structure for this new economics was completed by one of the preeminent minds of the twentieth century, Claude Elwood Shannon (1916–2001).

In a long career at MIT and AT&T's Bell Laboratories, Shannon was a man of toys, games, and surprises. His inventions all tended to be underestimated at first, only to later become resonant themes of his time and technology—from computer science and artificial intelligence to investment strategy and Internet architecture. As a boy during the roaring twenties in snowy northern Michigan, young Claude—grandson of a tinkering farmer who held a patent for a washing machine—made a telegraph line using the barbed-wire fence between his house and a friend's half a mile away. "Later," he said, "we scrounged telephone equipment from the local exchange and connected up a telephone." Thus he recapitulated the pivotal moment in the history of his later employer: from telegraph to telephone.

There is no record of what Shannon and the world would come to call the "channel capacity" of the fence. But later, Shannon's followers at industry conferences would ascribe a "Shannon capacity" of gigabits per second to barbed wire and joke about the "Shannon limit" of a long strand of linguini.

Shannon's contributions in telephony would follow his contributions in computing, all of which in turn were subsumed by higher abstractions in a theory of information. His award-winning master's thesis at MIT jump-started the computer age by demonstrating that the existing "relay" switching circuits from telephone exchanges could express the nineteenth-century algebra of logic George Boole invented, which became the prevailing logic of computing. A key insight came from an analogy with the game of twenty questions: paring down a complex problem to a chain of binary, yes-no choices, which Shannon may have been the first to dub "bits." Then this telephonic tinkerer went to work for Bell Labs at its creative height, when it was a place where a young genius could comfortably unicycle down the hallways juggling several balls over his head.

He worked on cryptography there during the war and talked about thinking machines over tea with the visiting British mathematician Alan Turing, whose conception of a generic abstract computer architecture made him, one can argue, the progenitor of information theory. At Bletchley Park in Britain, Turing's contributions to breaking German codes were critical to the Allied victory. During these wartime teas, the two computing-obsessed cryptographers also discussed what Shannon described as his burgeoning "notions on Information Theory" (for which Turing provided "a fair amount of negative feedback").

In 1948, Shannon published those notions in *The Bell System Technical Journal* as a seventy-eight-page monograph, "The Mathematical Theory of Communication." (The next year it reappeared as a book, with an introduction by Warren Weaver, one of America's leading wartime scientists.)[1] It became the central document of the dominant technology of the age, and it still resonates today as the theoretical underpinning for the Internet.

Shannon's first wife described the arresting magnetism of his countenance as "Christlike." Like Leonardo da Vinci and his fellow computing pioneer Charles Babbage, he was said by one purported witness to have built floating shoes for walking on water. With his second wife, herself a "computer" when he met her at AT&T, he created a home full of pianos, unicycles, chess-playing machines, and his own surprising congeries of seriously playful gadgets. These included a mechanical white mouse named Theseus—built soon after he wrote the information theory monograph—which could learn its way through a maze; a calculator that worked in Roman numerals; a rocket-powered Frisbee; a chair lift to take his children down to the nearby lake; a diorama in which three tiny clowns juggled eleven rings, ten balls, and seven clubs; and an analog computer and radio apparatus, built with the help of blackjack card-counter and fellow MIT professor Edward Thorp, to beat the roulette wheels at Las Vegas. (The apparatus worked in Shannon's basement but failed in the casino). Later an uncannily successful investor in technology stocks, Shannon insisted on the crucial differences between a casino and a stock exchange that eluded some of his followers.

When I wrote my book *Microcosm*, on the rise of the microchip, I was entranced with physics and was sure that the invention of the transistor at Bell Labs in 1948 was the paramount event of the postwar decade. Today, I find that physicists are entranced with the theory of information. I believe, with his biographer James Gleick, that Shannon's information theory was a breakthrough comparable to the transistor. While the transistor is ubiquitous today in information technology, Shannon's theories play a role in all the ascendant systems of the age. As universal principles, they grow more fertile as time passes. Every few weeks, I encounter another company whose work is rooted in Shannon's theories, full of earnest young engineers conspiring to beat the Shannon limit. Current technology seems to be both Shannon-limited and Shannon-enabled. So is the modern world.

Let us imagine the lineaments of an economics of disorder, disequilibrium, and surprise that could explain and measure the contributions of entrepreneurs. Such an economics would begin with the Smithian mold of order and equilibrium. Smith himself spoke of property rights, free trade, sound currency, and modest taxation as conditions necessary for prosperity. He was right: disorder, disequilibrium, chaos, and noise inhibit the creative acts that engender growth. The ultimate physical entropy envisaged as the heat death of the universe, in its total disorder, affords no room for invention or surprise. But entrepreneurial disorder is not chaos or mere noise. Entrepreneurial disorder is some combination of order and upheaval that might be termed "informative disorder."

Shannon defined information in terms of digital bits and measured it by the concept of *information entropy*: unexpected or surprising bits. The man who supposedly coined the term was John von Neumann, inventor of computer architectures, game theory, quantum math, nuclear devices, military strategies, and cellular automata, among other ingenious things. Encountering von Neumann in a corridor at MIT, Shannon allegedly told him about his new idea. Von Neumann suggested that he name it "entropy" after the thermodynamic concept. According to Shannon, von Neumann liked the term because no one knew what it meant.

Shannon's entropy is governed by a logarithmic equation nearly identical to the thermodynamic equation of Rudolf Clausius that describes physical entropy. But the parallels between the two entropies conceal several pitfalls for the unwary. Physical entropy is maximized when all the molecules in a physical system are at an equal temperature and thus cannot yield any more energy. Shannon's entropy is maximized when all the bits in a message are equally improbable and thus cannot be further compressed without loss of information. These two identical equations point to a deeper affinity that the physicist Seth Lloyd identifies as the foundation of all material reality—at the beginning was the entropic bit.[2]

For the purposes of economics, the key insight of information theory is that information is measured by the degree to which it is unexpected. Information is "news," gauged by its *surprisal*, which is the entropy. A stream of predictable bits conveys no information at all. A stream of uncoded chaotic noise conveys no information either.

In Shannon's scheme, a source selects a message from a portfolio of possible messages, encodes it by resorting to a dictionary or lookup table using a specified alphabet, and then transcribes the encoded message into a form that can be transmitted down a channel. Afflicting that channel is always some level of noise or interference. At the destination, the receiver decodes the message, translating it back into its original form. This is what is happening when a radio station modulates electromagnetic waves, and your car radio demodulates those waves, translating them back into the original sounds from the radio station.

Part of the genius of information theory is the understanding that this ordinary concept of communication through space extends also through time. A compact disk, iPod memory, or Tivo personal video recorder also conducts a transmission from a source (the original song or other content) through a channel (the CD, DVD, microchip memory, or "hard drive") to a receiver separated chiefly by time. In all these cases, the success of the transmission depends on the existence of a channel that does not change substantially during the course of the communication, either in space or in time.

Change in the channel is called noise, and an ideal channel is perfectly *linear*. What comes out is identical to what goes in. A good channel, whether for telephony, television, or data storage, does not change substantially during the period between the transmission and the receipt of the message. Because the channel is changeless, the message in the channel can communicate changes. The message of change can be distinguished from the unchanging parameters of the channel.

In that radio transmission, a voice or other acoustic signal is imposed on a band of electromagnetic waves through a modulation scheme. This set of rules allows a relatively high-frequency non-mechanical wave (measured in kilohertz to gigahertz and traveling at the speed of light) to carry a translated version of the desired sound, which the human ear can receive only in the form of a lower frequency mechanical wave (measured in acoustic hertz to low kilohertz and traveling close to a million times slower). The receiver can recover the modulation changes of amplitude or frequency or phase (timing) that encode the voice merely by subtracting the changeless radio waves. This process of recovery can occur years later if the modulated waves are sampled and stored on a disk or long term memory.

The great accomplishment in information theory was the development of a rigorous mathematical discipline to define and measure the information in the message sent down the channel. Shannon's entropy or surprisal defines and quantifies the information in a message. Like physical entropy, information entropy is always a positive number measured by minus the base two logarithm of its probability.

Information in Shannon's scheme is quantified in terms of a probability because Shannon interpreted the message as a selection or choice from a limited alphabet. Entropy is thus a measure of freedom of choice. In the simplest case of maximum entropy of equally probable elements, the uncertainty is merely the inverse of the number of elements or symbols. A coin toss offers two possibilities, heads or tails; the probability of either is one out of two; the logarithm of one half is *minus one*. With the minus canceled by Shannon's minus, a coin toss can yield *one* bit of information or surprisal. A series of bits of probability one out of two does not provide a 50-percent correct transmission. If it did, the communicator could replace the source with a random transmitter and get half the information right. The probability alone does not tell the receiver which bits are correct. It is the entropy that measures the information.

For another familiar example, the likelihood that any particular facet of a die turns up in a throw of dice is one-sixth, because there are six possibilities, all equally improbable. The communication power, though, is gauged not by its likelihood of one in six, but by the uncertainty resolved or dispersed by the message. One out of six is two to the minus 2.58, yielding an entropy or surprisal of 2.58 bits per throw.

Shannon's entropy gauged the surprisal of any communication that takes place over space or time. By quantifying the amount of information, he also was able to define both the capacity of a given channel for carrying information and the effect of noise on that carrying capacity.

From Shannon's information theory—his definition of the bit, his explanation and calculation of surprisal or entropy, his gauge of channel capacity, his profound explorations of the effect and nature of noise or interference, his abstract theory of cryptography, his projections for multi-user channels, his rules of redundancy and error correction, and his elaborate understanding of codes—would stem most of the technology of this information age.

Working at Bell Labs, Shannon focused on the concerns of the world's largest telephone company. But he offered cues for the application of his ideas in larger domains. His doctoral thesis in 1940 was titled "An Algebra for Theoretical Genetics." Armed with his later information theory insights, he included genetic transmissions as an example of communication over evolutionary time through the channel of the world. He estimated the total information complement in a human being's chromosomes to be hundreds of thousands of bits. Though he vastly underestimated the size of the genome, missing the current estimate of six billion bits by a factor of four thousand, he was nevertheless the first to assert that the human genetic inheritance consists of encoded information measurable in bits. By extending his theory to biological phenomena, he opened the door to its extension into economics, although to the end of his life in 2001 he remained cautious about the larger social applications of his mathematical concept.

It was Shannon's caution, his disciplined reluctance to contaminate his pure theory with wider concepts of semantic meaning and creative content, that made his formulations so generally applicable. Shannon did not create a science of any specific kind of communication. His science is not confined to telephone or television communications, or to physical transmission over radio waves or down wires, or to transmission of English language messages or numerical messages, or to the measurement of the properties of music or genomes or poems or political speeches or business letters. He did not supply a theory for communicating any particular language or code, though he was fascinated by measures of the redundancy of English.

Shannon offered a theory of messengers and messages, without a theory of the ultimate source of the message in a particular human mind with specific purposes, meanings, projects, goals, and philosophies. Because Shannon was remorselessly rigorous and restrained, his theory could be brought to bear on almost anything transmitted over time and space in the presence of noise or interference—including business ideas, entrepreneurial creations, economic profits, monetary currency values, private property protections, and innovative processes that impel economic growth.

An entrepreneur is the creator and manager of a business concept that he wishes to make a reality in time and space. Let us imagine Steve Jobs and the iPod. When he conceives the idea in his mind, he must then express, or "encode," it in a particular physical form that can be transmitted into a marketplace. This requires design, engineering, manufacturing, marketing, and distribution. It is a complex endeavor dense with information at every stage.

As an entrepreneur and the CEO of Apple, Jobs controls many of the stages. But the ultimate success of the project depends on the existence of a channel through which it can be consummated over nearly a decade, while many other companies outside his control produce multifarious competitive or complementary creations. Vital to Apple's wireless achievements are advances in ceramic and plastic packaging, digital signal processing, radio communications, miniaturization of hard disks, non-volatile "flash" silicon memories, digital compression codes, and innumerable other technologies feeding an unfathomably long and roundabout chain of interdependent creations.

In biology itself, chemical and physical laws define many of the enabling regularities of the channel of the world. In the world of economics in which Jobs operated, he needs the stable existence of a "channel" that can enable the idea he conceives at one point in time and space to arrive at another point years later. Essential to the channel is the existence of the Smithian *order*. Jobs must be sure that the essential features of the economic system that is in place at the beginning of the process are still there at the end. Adam Smith defined those essential features of the channel as free trade, reasonable regulations, sound currencies, modest taxation, and reliable protection of property rights. No one has improved much on this list.

In other words, the entrepreneur needs a channel that in these critical respects does not drastically change. Technology can radically change, but the characteristics of the basic channel for free entrepreneurial creativity cannot change substantially. A sharp rise in tax rates, or laws against the ownership of rights to music, or regulations gravely inhibiting international trade would have impeded the channel for the iPod.

One fundamental principle of information theory distills all these considerations: the transmission of a high-entropy, surprising product requires a low-entropy, unsurprising channel largely free of interference. Interference can come from many sources. Acts of God like tsunamis and hurricanes have been known to do the job, though otherwise vigorous economies quickly recover from these disasters. For a particular entrepreneurial idea, interference may come in the form of a more powerful competing technology.

The most common and destructive source of noise, however, is precisely the institution on which we most depend to provide a clear and stable channel in the first place. When government either neglects its role as guardian of the channel or, worse, tries to help by becoming a transmitter and turning up the power on certain favored signals, the noise can be deafening.

A friendly government that excluded all Jobs's rivals from the channel or granted Jobs a monopoly on the distribution of music might have benefited Jobs for one product. But a ban on competitive products would thwart the necessary technological advances on which Jobs's future products would depend, and a high-entropy, government-dominated channel, full of unpredictable political interference and noise, would depress the sacrificial long-term investment of capital.

An entrepreneur contemplating his invention and its prospects for success must estimate its potential profitability. Profit is the name that economics assigns to the yield of investments. The levels of interest rates and their time- and risk-structures express the average yield across an entire economy, reflecting the existing pattern of production and expected values of currencies. Interest rates will define the opportunity cost of investments in new products: what other opportunities are missed on average as a result of pursuing one in particular.

Interest rates are critical for information-theory economic analysis because they are an index of real economic conditions. If the government manipulates them, they will issue false signals, breeding confusion that undermines entrepreneurial activity. For example, if the government keeps interest rates artificially low for institutions that finance it—as it has been doing in the United States—the channel is seriously distorted. The interest rates are noise rather than signal. Interest rates near zero cause finance to hypertrophy as privileged borrowers reinvest government funds in government securities. Only a small portion of these funds goes to useful "infrastructure," while the rest is burned off in consumption beyond our means.

An entrepreneur making large outlays to bring a major product to market over a number of years will normally have to promise a profit, perhaps to venture capitalists or a board of directors, far exceeding the interest rate. The economy at large does not expect this entrepreneurial profit. The large, established companies that dominate the marketplace do not anticipate it. Profits differentiate between the normally predictable yield of commodities and the unexpected returns of creativity. The entrepreneur's new product or business surprises the economy, and his payback will be surprising—it will disrupt the equilibrium of the existing order. If established companies can manipulate the channel to protect their own products, businesses, and margins, a new product cannot pass through.

The unexpected financial profit is surprisal or entropy—what Peter Drucker called an "upside surprise."[3] Drucker pointed out that most measured financial "profits" are not real in this sense. They merely cover the cost of capital—the return of interest. Innovation is the source of real profit, entropic profit, which derives from the upside surprises of entrepreneurial creativity.

In order for the entrepreneur to succeed, he must know that, if his creation generates an upside surprise, the related profits will not be confiscated or taxed away. If they may be confiscated, his entire project will not be able to attract the necessary resources to bring it to market.

Linking innovation, surprise, and profit, Shannon's entropy is the heart of the economics of information theory. Signaling the arrival of an invention or disruptive innovation is first its surprisal, then its yield beyond the interest rate—its profit, a further form of Shannon's entropy. As the market absorbs a new product, however, its entropy declines until its margins converge with prevailing risk-adjusted interest rates. The entrepreneur must then move on to new surprises.

The economics of entropy describe the process by which the entrepreneur translates the idea in his imagination into a practical form. In those visionary realms, entropy is essentially infinite and unconstrained and thus irrelevant to economic models. But to make what he has imagined practical, the entrepreneur must make specific choices among existing resources and strategic possibilities. Entropy here signifies his freedom of choice.

As Shannon understood, the creation process itself defies every logical and mathematical system. It springs not from secure knowledge but from falsifiable tests of commercial hypotheses. It is not an expression of past knowledge but of the fertility of consciousness, will, discipline, imagination, and art.

Like all logical systems founded on mathematical reasoning, information theory depends on axioms that it cannot prove. These axioms comprise the content flowing through the conduits of the economy, and they come from the minds of creators endowed with freedom of choice. Once the entrepreneur turns his idea into a reality, projecting it into the channel of the economy as a falsifiable experiment, it falls into the Shannon scheme. Measured by their entropy—their content and surprisal—new products face the test of the market that they create and the profits they engender.

Entropy
Economics

THROUGH WARS AND DEPRESSIONS, through booms and benisons of prosperity, the central failure of economics has been its inability to grasp the centrality of entrepreneurial creation in economic life.

The key force of economic advance is the entrepreneur, who on his own, without governmental cues or expert consultation or even a defined market, creates new goods, services, business plans, and projects. Economic growth and progress, jobs and welfare, markets and demand all stem from this creativity of the entrepreneur. Population growth, capital accumulation, economic efficiency, and even scientific advances are all less important than entrepreneurial creativity. And governmental interventions in the economy are distractions—"noise on the line"—that nearly always retard expansion. Failing to see the centrality of entrepreneurial creativity, economists everywhere have counseled governments to attend to the money supply, aggregate demand, consumer confidence, trade imbalances, budget deficits, capital flows—to attend to everything except what matters most: the environment for innovation.

The failure of nearly all schools of economics has its roots in the canonical works of Adam Smith. As Bonnie Prince Charlie, his fellow Scot, angled for the British throne, and slave ships plied their trade across the Middle Passage, Smith—an archetypical absent-minded professor of what was then called "moral philosophy" and one of the leaders of the Enlightenment in Scotland—retired to his mother's house north of Edinburgh to write. He produced *An Inquiry into the Nature and Causes of the Wealth of Nations*, published ten years later in the notable year of 1776. Inspired by the dazzling physical "System of the World" unveiled by Isaac Newton in the previous century, Smith addressed himself to the challenge of understanding the existing business practices and organizations that were then beginning to make the world rapidly richer.

By examining these *systems*, Smith invented the economics and business philosophy that governs the modern imagination, imprinting it with indelible images of mass production in a pin factory, comparative advantages in international trade, and a ubiquitous "distribution" of goods and incomes stemming from the increasingly elaborate "division of labor." *The Wealth of Nations* depicts macroeconomics as a "Great Machine" in which every cog of every gear, governed by an "invisible hand," functions perfectly in its time and place, as smoothly and reliably as Newton's gravity. There were entrepreneurs, to be sure, but they "seldom meet together" without the conversation ending "in a conspiracy against the public, or in some contrivance to raise prices."

In the end, Smith was the greatest defender of free enterprise and open systems and untrammeled trade between free peoples. But in his system, entrepreneurial creation is subsumed under the rubric of the "division of labor"—the extent of which, so he ordained, "is determined by the extent of the market." That there is no market without entrepreneurs passed unnoticed in the two centuries that followed Smith's work. Even the Austrian titan Friedrich Hayek missed it in his valiant defense of free markets after World War II. Among advocates of free markets, Hayek's "spontaneous order" remains the prevailing image of economic organization.

Amid all that spontaneous order and economic equilibrium, however, entrepreneurs and their creations continued to crop up disruptively. Beginning with Smith, economists acknowledged their importance. But nearly every leading economic theorist from Smith to Sigmund Sismondi, from Max Weber to Karl Marx, from Joseph Schumpeter to Frank Knight, from John Kenneth Galbraith to Paul Samuelson—whether a friend or foe of the free market—predicted the exhaustion and demise of the entrepreneurial role. Entrepreneurs might have their day and from time to time would win great glory, but with the increasing spread of the market (driven by trade agreements) and the ever-developing division of labor (driven by the expansion of the market), the entrepreneur would recede. Dominating the new economy would be giant institutions, which politics and scientific expertise would buttress, and which presumably would be too big to fail and too powerful to be challenged.

Capitalism would evolve, even its proponents said, into a "stationary state" where abundance ruled, eclipsing the role of individual entrepreneurs reaping profits from scarcity. More modern theorists find free enterprise's rough edges intolerable in light of the fragility of the environment. All these economists deem entrepreneurial leadership as a transitory and dispensable stage of capitalism, ripe for hierarchical regimes of expertise and specialization to supplant. At some point, they believe, the surprises will stop.

This attitude toward enterprise, which puts the equilibrium of markets first and looks forward to disruptive innovators withering away, has produced the current economic crisis and the crisis of economic thought. For contrary to every economic theory of a stationary state, a new industrial state, or an eclipse of capital, the role of entrepreneurial risk-taking is more important than ever before. In 2010, companies launched by entrepreneurs backed by venture capital generated over a fifth of America's gross domestic product. In that same year, companies less than five years old created all of the new jobs. (The older ones actually shed jobs.)

Men such as Peter Drucker,[1] Daniel Bell,[2] Marshall McLuhan,[3] Alvin Toffler,[4] Stewart Brand,[5] and John Perry Barlow[6] wrote more presciently than they knew when they called this era an age of information and celebrated an information economy. They were intrigued by the extent to which economic activity, from finance and insurance to international trade and education, from computers and communications to managerial consulting and biotechnology, was largely devoted to the creation and processing of symbols, software, programming languages, and networks. These information products, however, although powerful signs and symbols of a new economics, are not its substance.

The most important feature of an information economy, in which information is defined as surprise, is the overthrow, not the attainment, of equilibrium. The science that we have come to know as information theory establishes the supremacy of the entrepreneur because it appreciates the powerful connection between destruction and what Schumpeter described as "creative destruction," between chaos and creativity.

Shannon stressed the stochastic nature of the messages that information theory follows. "Stochastic" comes from a Greek word that means "to aim at," combining probabilities with skill, randomness with structure. As adopted by physicists, the word applies to random processes such as Brownian motion (the ceaseless agitation of thermal molecules) that operate unpredictably within a predictable regime of physical and chemical laws. A stochastic process is neither determined nor utterly random. The stream of high-entropy bits in a message flows through a low-entropy, determined channel. In other words, it is bounded noise. The ultimate thermodynamic entropy of heat death is all noise and no carrier. The ultimate Shannon entropy is noise-like, apparently random flow—that "white noise" that Qualcomm's Viterbi celebrated—in a relatively noiseless, structured vessel.

Any observer can identify as a signal a repeated pulse in a clear channel. Like a beat on a heart-rate monitor, it is a recognizable low-entropy carrier. Unless the heartbeat becomes irregular, little information is conveyed. A steady pulse is what is expected. By contrast, a channel operating at its Shannon limit will contain so much surprise—so much apparent randomness—that it will actually appear as random noise to any recipient unequipped with the proper decoding device.

The epitome of a low-entropy carrier is the electromagnetic spectrum, from radio waves to light waves, a regular radiance of perfect sine waves governed by the unchanging speed of light. Electromagnetic waves are differentiated only by their hertz, the number of times they undulate every second. Because of its supreme regularity, the spectrum can carry a measurable modulation, a deliberate distortion, which is detectable at a remote receiver. The electromagnetic spectrum, therefore, is perfectly suited for transmitting the entropy of the information economy, whether through wires or through the air.

Having studied with Shannon at MIT, Qualcomm's founders were far better prepared for the information economy than were the veterans of analog radar, telephony, and television who dominated most other wireless companies. Qualcomm is *based* on a theory of information, a theory so fundamental that it pushes Qualcomm into all contiguous markets—microchips, computers, networks, and software—and makes Qualcomm an exemplary company of the information age.

In particular, Irwin Jacobs and Andrew Viterbi grasped Shannon's revelation of the relationship between information, power, and noise in digital systems. What matters for digital communications is not *maximizing* power. Large variations in power constitute noise. The goal is *controlling* power and reducing it to the minimum at which the on-off bits are intelligible. The object of digital wireless is not to blast a particular signal, like Rush Limbaugh's voice vibrations, on fifty-kilowatt waves all across the fruited plain. It is to maximize the number of unexpected bits—the amount of entropy—that all radios can transmit and receive in a particular cell. Operating at watts rather than kilowatts, Qualcomm was the first wireless company designed for the Shannon era of low-power communications.

Viterbi's low-powered white noise is the secret. White noise is defined by its randomness. Each sound or signal, independent of previous signals, is utterly unpredictable; each bit is unexpected. That's what makes pure noise. In information theory, it is in principle impossible to differentiate such random noise from a series of unrelated creative surprises. Both are gauged by their entropy or surprisal, and both seem random. Unless you have the code, they both look the same.

This principle of information theory stultifies all the theories that try to reduce entrepreneurial creativity to "random walks" and "fractal" markets and black swan spikes and singularities based on looking at an oscilloscopic rendition of market prices and movements. The world is full of noise that looks random, but the stock market or Silicon Valley is not random, despite the appearance of an unpredictable path of prices or company ups and downs. Shannon, a shrewd investor himself, understood this point. He summed it up this way in 1987, responding to the efficient markets theory: "We do study the graphs and the charts. The bottom line is that the mathematics is not as important in my opinion as the people and the product."[7] Qualcomm's rise was not random but creative. It was an expression of entropy as freedom of creative choice. In order to understand the movement of prices, you need not an oscilloscope to measure the entire market and reduce it to noise, but a microscope to investigate the creative process behind every company and its price.

Entrepreneurship is devoted to creation of new goods and services. Creativity is always surprising. That is why it cannot be planned or demanded by governments or even by customers. As Steve Jobs put it, explaining his contempt for market surveys, "It's really hard to design products by focus groups. A lot of times, people don't know what they want until you show it to them."[8] As Henry Ford said many years earlier: "If I had listened to my customers, I would have built a faster horse."[9] Inventions in general express Shannon entropy. They come from the supply side.

Qualcomm's CDMA was a perfect high-entropy supply-side surprise. Experts like Lusignan actually believed the Qualcomm system was impossible to implement successfully. No one in the industry thought he wanted it. No one demanded it, and many industry leaders fought to stop it. But everyone in the industry ended up using it.

Capturing the unexpected return beyond the predicted return—i.e., the interest rate—were Qualcomm's profits, the measure of the entropy in its system. Profits and losses are the unexpected results beyond the interest rate. Entropy registers when the cup overflows or empties unexpectedly.

One of the world's most creative companies, Qualcomm has also, for close to two decades, been among the world's most profitable companies. Based on a theory of entropy, it led its industry in profits, another form of entropy. In recent years, Qualcomm's stock was excelled by Apple's, which followed Steve Jobs's supply-side inspiration to become the most profitable and most highly valued large company in the world. All Apple's wireless technologies and design innovations stemmed from the same information theory that propelled Qualcomm.

Viterbi's white noise also clarifies economic policy. White noise implies a multitude of independent initiatives, each one different and none dominant, collectively maximizing the entropy of the system. In economic policy, this describes a proliferation of small business startups and creative experiments, collectively maximizing the profits of innovation and reflecting the dispersal of human knowledge.

On the other hand, we have the high-powered analog signals that Lusignan and most of the rest of Silicon Valley thought were the best means to maximize the flow of entropy. These signals are analogous to gigantic corporations, government programs, and banks that use their power to be certified as too big to fail.

Each of these strategies can work for a while. With inspired leadership, large corporations and government bodies can function effectively for a limited period. In wartime, with clear objectives for the entire economy, large companies can mobilize to deliver complex military goods. Most new systems, however, depend on technology from startups. Like all analog transmissions, large corporations are ultimately corruptible by the centripetal pull of power, of interference, and of agency conflicts. Only the white-noise strategy allows the correction of redundancy and error among the multitude of creative signals. It ends with a more robust economy, more innovation, and more jobs.

The superiority of multifarious entrepreneurship finds strong support from the U.S. Census Bureau's Business Dynamics Statistics program. Between 1996 and 2009, the data show, virtually all the new jobs came from startups. Even in 2009, in the midst of depression while older and larger firms were shedding some seven million jobs, new companies added 2.3 million jobs. In 2010, 21 percent of U.S. GDP and more than 60 percent of stock market share value came from companies launched with the help of the venture capital system. The conclusion to be drawn is that essentially all net new jobs come from fast-rising startups, especially technology firms backed by venture capital. This is the heart of the economy. The supply-side solution to our current economic stagnation is a return to the low-entropy carrier: predictable rules of taxation, regulation, immigration, and monetary stability, which favor long-term investments in innovative new companies.

These startups drive all innovation and employment growth. Doing new things, they are devoted to effectiveness, not efficiency. This dominance of the innovators is exactly what the theory of entropy would lead you to expect in a capitalist economy.

Viterbi's view of the affinity between noise and information, entropy and communication, baffles economists and theorists of capitalism. Ever since *The Wealth of Nations*, economists have imagined that entrepreneurs seek equilibrium and order. Hundreds of conservative economists have followed Friedrich Hayek into the intellectual swamp of "spontaneous order" and self-organization. On the farther shores of libertarianism, belief in spontaneous order leads to the indulgence of anarchy, prompting theorists such as Murray Rothbard to regard all government as unnecessary tyranny. Rothbard was the economist behind the conspiratorial ruminations of Ron Paul, seeing no essential difference between the American, Israeli, and Iranian governments.

Entropy is a measure of surprise, disorder, randomness, noise, disequilibrium, and complexity. It is a measure of freedom of choice. Its economic fruits are creativity and profit. Its opposites are predictability, order, low complexity, determinism, equilibrium, and tyranny.

Predictability and order are *not* spontaneous and cannot be left to an invisible hand. It takes a low-entropy carrier (no surprises) to bear high-entropy information (full of surprisal). In capitalism, the predictable carriers are the rule of law, the maintenance of order, the defense of property rights, the reliability and restraint of regulation, the transparency of accounts, the stability of money, the discipline and futurity of family life, and a level of taxation commensurate with a modest and predictable role of government.

These low-entropy carriers do not emerge spontaneously. They are the effects of political leadership and sacrifice, prudence and forbearance, wisdom and courage. Sometimes they must be defended by military force. They originated historically in a religious faith in the transcendent order of the universe. They embody a hierarchic principle. It is these low-entropy carriers that enable the high-entropy creations of successful capitalism.

Romney, Bain, and the Curve of Learning

IN THE EARLY 1980s, soon after the publication of my book *Wealth and Poverty*, I received a phone call from a man with a mellow mid-Southern accent, honed from roots in rural Tennessee and long years talking down to top executives of American enterprise. William Bain was the name, "but everyone calls me Bill," he said. I had never heard of him or his eponymous company, a consulting firm in Boston with ambitions in venture capital, but I was soon on board, calling him "Bill" and listening closely to what he had to say.

He invited me to speak to his team of Bain & Company partners, and also, if he might, ... no offense... he wanted to impart some ideas of his own, some points I might have missed on supply-side economics. "We've done some research," he said, "that shows the theory is much more general and powerful than even you believe."

As one unused to charges of underestimating the power of supply-side theory, I was intrigued. I went on to give many hundreds of speeches and participate in scores of debates with such figures as Robert Reich, the

Harvard professor who became secretary of labor under President Bill Clinton; Lester Thurow, the eminent MIT professor and bestselling author; and the legendary six-foot-eight-inch tribune of tall taxes, John Kenneth Galbraith of Harvard and Gstaad. I received astonishing fees of up to six figures (a level I breached for an event in Cambridge, England, for a German bank). At the time, supply-side was that hot. But every speech and book that I produced from then on bore the imprint of my conversations with people at Bain. I learned more from them than from any other audience and, truth be told, from my four years at Harvard, which included little economics beyond disgruntled attendance at a lecture by Galbraith, who read word-for-word from his bestselling book.

What I learned from Bain was the key role of learning and information in economics. Long before I immersed myself in the works of Claude Shannon, the practical lesson in information economics that I gained with Bain thirty years ago was transformative. It replaced the stimulus-and-response incentive structure of original supply-side theory with the sound microeconomic underpinnings of learning, information, and entrepreneurial surprise.

Across from the oldest cemetery in Boston, under the chandeliers of the Parker House, the Bain meeting brought together, as I recall, perhaps a hundred men in suits. Notable in the group were Bain himself, a spruce young man with blond hair brushed straight back, and an Israeli woman, Orit Gadiesh, who within ten years would rise to the top of the organization. But Bain seemed more eager to introduce me to one Mitt Romney.

I recall the introduction vividly, and not just because of a dismal year I had spent working for his father, George, then the governor of Michigan, in 1966 and 1967, ghost-writing a never-published presidential campaign tract called *The Mission and the Dream*. I had been a speechwriter for Nelson Rockefeller, a dyslexic who treated his writers as royalty. Romney treated my emergent tome with all the gravity he might have devoted to the owner's manual for the Nash Rambler. I ultimately learned from my time in Lansing that the worst times of your life can be redemptive. (I would later learn, as a dot-com-era businessman, that the euphoric times can be catastrophic.)

My loss of a pay check with the collapse of George Romney's campaign impeded my efforts to date the damsels of Michigan State and left me subsisting on ever more dubious items from deep in my motel refrigerator. My ordeal was not improved by my residual link to the Rockefeller campaign, embodied in a borrowed white Plymouth whose parking tickets on the street in front of the Romney office—which I ignored—went to 5600 Rockefeller Plaza in New York, followed in the course of time by two solemn state policemen, who travelled all the way to Rockefeller Center and up to the Rockefeller offices to collect the money. The dutiful Michigan troopers confronted my baffled godfather, David Rockefeller, in his office (his daughter Peggy had lent me the car). This pilgrimage marked a low point in my relations with both the Eastern Establishment and the Romney campaign.

Nonetheless the research and themes of *The Mission and the Dream*—worked out during dark months in a Lansing motel with only occasional interaction with the putative author, whom a coterie of careful protectors guarded from any haunting by ghosts from east of Grosse Pointe—laid the groundwork for *Wealth and Poverty*. Romney's failure to pay me meant that the work I had expended on *The Mission and The Dream* was mine to keep. A decade or so later, with a rapidity that would have been impossible starting from scratch, I could turn it into the book that popularized supply-side economics and made me Ronald Reagan's most-quoted living author.

I remembered Mitt at the Parker House because of his striking looks, confidence, and charisma, which I remarked at the time were even more impressive than his father's. With degrees from Harvard's business and law schools, he had been one of the leading candidates in the super-competitive realms of the Boston consultancies. More importantly, I believe, he conveyed the gravitas of a graduate of the Mormon school of hard knocks and hair shirts as a missionary for two years selling his religion's bread-and-water regimen in Bordeaux, France. He had also been a narrow survivor of a fatal head-on collision with the careening car of a local Catholic priest, which took the life of the wife of Romney's mission president.[1]

Consultants with a modus operandi—unique to Bain—of attaching themselves to companies only at the CEO level, none of these young corporate quarterbacks showed any propensity for attentive service on the benches of life. These were young men with the leverage and audacity to advise famously imperious chief executives on life-and-death matters in their companies. At the Parker House meeting in 1981, among all these luminaries, Mitt was already ascendant.

He seemed only mildly interested in my association with his father George, and he gave me the impression that I had lived in Lansing longer than he had. Out in Lansing, I had come to admire George as a man who got up every morning at five and played a round of golf with three golf balls in parallel, while running from shot to shot. He was a magnificent creature, an inspirational entrepreneurial leader at American Motors, and an exemplary father for Mitt. But I eventually found him gullible to the point of brainwashing about liberal ideas.

The elder Romney was abashed by Ivy-League expertise, the great peril of establishment Republicans from the time of both Bushes through the presidential candidacy of John McCain. All cherish the illusion that leading Yale, Harvard, and Princeton economists possess vital wisdom about the economy. They generally don't. Their preoccupation with static macroeconomic data blinds them to the actual life and dynamics of entrepreneurship. Their preoccupation with liabilities and debt blinds them to the impact of their policies on the value of economic assets. Their GDP model, where everything is measured as a kind of spending—power rather than knowledge—pushes them to manipulative policies and redistribution inimical to business value and growth. Believing that a weaker dollar is just the thing to spur a sluggish economy, by hyping the spending category of "net exports," they miss the consequent devaluation of all the assets of the country.

George Romney capitulated to these forces. His great achievement as a big-spending governor of Michigan was the enactment of a state income tax. As Nixon's secretary of housing and urban development, he followed the liberal temptation into a series of ineffectual big-government programs. Tangentially implicated in a 1972 scandal involving mortgage-backed securities from the Federal Housing Administration and Ginnie Mae, he could even be described as an early source of the feel-good finance of confectionary home ownership that eventually brought down the economy at the end of the Bush years.

When Mitt Romney moved into politics, I hoped that the son would excel the father as a man with a mind of his own, resistant to wishy-washy "compassionate conservatism," the Republican form of Obama's hope-and-change Marxism. Thus I looked on with nothing short of horror at his can-you-top-this effort to win a Massachusetts senate race by sloughing off every principle of his upbringing to thread his way down the slim sidelines to the left of Ted Kennedy. Stepping well out of bounds, he even repudiated Reagan, who had actually won Massachusetts twice.

On the surface, Romney was an improbable conservative champion. But apart from his round-heeled political compromises in Massachusetts, I knew he combined great abilities as a leader with a grasp of supply-side theory that surpassed that of other Republicans. After spinning Bain Capital out of Bain & Company in 1984, he compiled an astonishing record in private equity that was never effectively impeached despite his political foes' furious efforts. He capped this run with a bold and ruthless rescue of the parent firm when Bain & Company ran off the rails as the founders attempted to cash out in the early 1990s. Romney cut the founders' share by half; slashed compensation; fired half the people; shook or faced down the creditors, including Goldman Sachs and the U.S. government; and saved the day for his future entry into politics. He doesn't boast about it, but this flawless performance in the clutch expressed Romney's extraordinary gift for crisis management, demon-strated again and again in his career and personal life.

Back in the 1980s, Bain Capital under Romney was a spearhead of a massive national movement of corporate restructuring. The high tax rates of the inflationary 1970s had provoked a wave of deadening con-glomeration and corporate bloat that resulted in a catastrophic 60-per-cent decline in the real value of corporate equity. This was the era of palatial new corporate headquarters, jet fleets, and lavish entertainment budgets, all serving incoherent jumbles of unrelated companies whose equity was worth less than the sum of their parts. "Splurge or merge" was the order of the day as corporations sought to avoid the confiscation of their profits through inflation and taxes, which could rise to effective rates above 100 percent of real returns.

Conglomerates artfully combined companies nursing losses with companies harvesting profits, buffering the impact of the deadly tax regime. Then Ronald Reagan's counter-inflationary supply-side tax policies, coupled with Paul Volcker's monetary contraction, rendered these combinations dysfunctional. They had to be dismantled and reorganized for a low-tax, low-inflation regime, despite intense internal opposition. Romney was a leader of this restructuring campaign, which radically increased the value of U.S. businesses.

This history is deeply relevant today. For the first time since the Carter years, the U.S. once again leads the world in corporate tax rates and importunate lobbying for government succor. Independent businesses are foundering, and the country is suffering a net flight of capital and skilled manpower abroad. It is the resulting collapse of America's assets that makes its liabilities increasingly unsupportable.

Despite the desperate need for restructuring of both government and business, our national memory is clouded. Attempting to show that Bain Capital threw people out of work, Romeny's critics in the media focused relentlessly on anecdotes from particular company turnarounds. These single-company narratives, to which Romney's defenders also resort, are nearly irrelevant to the economic lessons of the general restructuring that released capital for better uses and more jobs and higher valuations all across the economy. One of the chief beneficiaries of the hundreds of billions of released and recycled money was venture capital, funding new technology startups such as Cisco and Google, which today collectively account for some 21 percent of GDP and more than 60 percent of the value of the nation's equities.

In the mid-1990s, Harvard Business School's Michael Jensen and his team closely studied the question of restructuring and job losses. Jensen calculated that between 1976 and 1993, a period that covers Romney's Bain Capital years, U.S. corporations conducted 42,621 mergers and acquisitions worth a total of $3.1 trillion. Selling firms won premiums of

41 percent, generating $899 billion in constant-dollar gains for share-holders (well over a trillion in today's dollars). Buying firms also gained on average, by increments that increased over the years.[2] Since Bain outperformed under Romney, its results were even better.

Lawrence Summers of the Harvard economics department (and later secretary of the Treasury) contended that these gains disguised wealth transfers from bondholders, workers, suppliers, and communities. Jensen disproved this charge, showing that in the aftermath of the transactions, there were sharp increases in capital expenditures, research and develop-ment, employment, and share value. During this period, encompassing the Reagan years, the United States easily led the world in job creation with between fifty and fifty-five million new jobs, at steadily rising pay, against only ten to fifteen million jobs lost. Since the United States was generating jobs far faster than its overseas rivals, this restructuring could hardly have caused job losses to foreign countries. America continued to lead the world in job creation, launched the computer revolution, and maintained its manufacturing employment level until the crash of 2000.

What accounts for the huge influence of private equity investments and buyouts? Jensen stresses the importance of realigning management with stockholders and overcoming the "agency problem." Ownership fosters good management because owners are their own agents. All other arrangements foster subtle or even open conflict between managers—who are tempted to enrich themselves, hire cronies, and build empires—and shareholders, who in general single-mindedly seek to maximize the value of the enterprise.

The underlying reason for the efficacy of private equity transactions, however—as I learned from my meetings with Bain—is the better align-ment of knowledge and power. Romney could build value for his investors because he combined the financial power of Bain with an intimate knowl-edge of all the companies in his portfolio and an understanding of the constantly changing economic landscape. Like Warren Buffett, John Doerr, and other successful investors who deal in entire companies, he exploited the legality of insider knowledge for owners and aspiring owners.

By contrast, "fair disclosure" securities laws sterilize information by channeling it through a public company's legal and public relations departments, denying shareholders access to the genuinely useful information about their own companies and trivializing their ownership. The perverse result of such laws is that when a company goes public, its information goes private. Lawyers and PR experts strain out all substance beyond the quarterly disclosure of enigmatic numbers.

"Holding companies" such as Berkshire Hathaway or General Electric, venture capitalists such as Doerr's Kleiner Perkins, and private equity players such as Romney's Bain Capital all escape this trap. They can intimately understand the investments they make. They legally join the knowledge of ownership with the power of profits. They are entrepreneurs, commanding the most powerful money in any economy: fully informed finance.

If Romney had been listening more attentively when I gave my speech back in 1982, he might have been more cogent in responding to the charges of "vulture capitalism" thirty years later. I followed the entrepreneurial economist Mark Skousen in showing that consumer spending is nowhere near 70 percent of the real economy (GDP leaves out all intermediate transactions in the supply chain) and is nearly irrelevant to economic growth ("supply creates its own demand").[3] I spoke on the centrality of venture capital and the power of entrepreneurs responding to tax-rate reductions: "High tax rates don't stop rich people from being rich; they stop everyone else from getting rich," I said. "Progressive tax rates don't redistribute incomes, they redistribute taxpayers... from factories and offices and onto foreign beaches and early retirements." These lines, old favorites, ring true today as we consider Europe or scrutinize California. And I made my case that capitalists thrive only by serving others. But in the early 1980s, I still did not really grasp the deeper sources of the effectiveness of venture capitalists and private equity players.

After my speech, I began an educational gauntlet at Bain, which set my course for years to come. My chief instructors were Bill Bain himself and another of his leading consultants with a political background and commanding personal presence, T. Coleman Andrews. The horn-rimmed black-locked namesake of his grandfather, who had run for president on the Dixiecrat ticket, Andrews smacked less of his Virginia heritage than of his years at Dartmouth, then leading the world in computer education under President John Kemeny. The young Virginian was Romney's chief companion on the road to raising funds for Bain.

Bain and Andrews explained to me that tax-rate reductions were just a special case of the strategy of aggressive price cutting on which Bain had based much of its consulting practice. Bain itself would often do its first project for free. "We have discovered," Bain said, "that aggressive price cuts can trigger a cascade of strategic benefits, not just expanding market share, building asset values, and increasing revenues and profits, but also gaining more knowledge of the strategic environment and provoking overreactions and blunders by rivals."

"Companies in trouble that raise their prices, on the other hand," Bain explained, "all too often begin a spiral of decline." The market darkens before them as they retreat from it into highly paid niches. Their technological progress slows as their volumes decline and rivals rush ahead into the future. Bain saw the United States under President Carter as a company in trouble that was raising its prices in response, with all the predictable bad effects, such as competitive losses to Japan and Germany, lower real revenues for the government, collapsing equity values and the famous "national malaise." The pattern is being repeated today under Obama.

At the heart of Bain's analysis was a proposition that originated in studies by research teams in the U.S. Navy during World War II—the famous *learning curve*. Bruce Henderson, another Tennessean with a Vanderbilt degree, led the Boston Consulting Group after the war in further development and extension of this seminal insight. In the 1960s, Henderson hired Bain and later Romney and his friend Benjamin Netanyahu, the future prime minister of Israel. All three gained their original grasp of capitalist dynamics at BCG, founding a theory of business economics on the intricacies of learning and innovation.

Generalized as the "experience curve," the theory holds that—largely because of on-the-job learning, broadly considered—the cost of producing any good or service declines by between 20 percent and 30 percent with every doubling of units sold. Growing apace with output and sales is entrepreneurial *knowledge*, which springs from improvements in every facet of the company; every manufacturing process; every detail of design, marketing, and management. Crucially, the curve extends to customers, who learn how to use the product and multiply applications for it as it drops in price.

The root of a company's value, then, is *experience*, which is a product of the knowledge in the company, the learning of the customers, the balance sheet of assets and liabilities, and, not least, the policy environment. Private equity investment firms like Bain Capital work on all these levels to achieve capital gains.

The experience curve charts the increase in prowess with experience and scale in the provision of any product, from pins to cookies, insurance policies to phone calls, pork bellies to chicken broilers, steel ingots to airplanes. But if you raise prices—or taxes—you slow down all this learning and experience, increasing average costs across a business and an economy, depleting asset values, exacerbating liabilities, and lowering both private profits and public revenues. Government becomes more powerful in relation to the private sector, although weaker in relation to the world.

Michael Rothschild, another business consultant and the author of *Bionomics,* soon detected the same learning curve throughout the biological world.[4] Rothschild showed that Henderson's 25-percent increase in efficiency with every doubling of accumulated output applies to everything from hominids hunting and gathering on an African savannah to rain-forest slime molds collecting nutrients. (Today slime molds are being tried as a material for biocomputing). *Bionomics* included the devastating observation that, in the crucial test of relevant intelligence, public-sector unions learn more slowly than tapeworms (parasites that only rarely devour their hosts).

A decade later, in the tour de force *The Singularity is Near*, the inventor and prophet Ray Kurzweil found exponential advances across the technoscape that dwarf the incremental advance of GDP cherished by economic histories.[5] Rothschild and Kurzweil had the gist of the story. Their nineteenth-century precursor Henry Adams arguably captured its essence in his "Law of Acceleration" in the *Education of Henry Adams*.[6] But Henderson and Bain were the first to explore in depth the sources of the curve in the economics of information and learning.

Romney, however, occasionally failed to grasp the significance of the curve for policy. Famously "data driven," he introduced a universal health-care bill in Massachusetts that worked perfectly on paper for the existing pattern of patients and facilities when it was passed. Things changed as the program went into effect. As Bain would have predicted, people learned, politicians took over, knowledge and power diverged, and the state's medical costs ended up doubling.

A decisive rule of social science long taught at Bain is that people learn how to exploit any good they experience as free. If it is paid for by taxes, it is tantamount to free for the user. While to the demand-side analyst, for example, free universal health insurance is a solution, to a supply side analyst it will be the problem because it thwarts learning and technological progress. Regardless of what politicians promise to gullible voters, government services cannot escape the constraints of supply and demand. A price of "free" evokes unbounded demand while choking off supply. A remote misbegotten cousin of Romneycare, for example, is Obamacare. It provides for 16,000 new IRS agents and new taxes and fees galore to fund it, but it actually degrades the power of physicians and restricts the supply of medical instruments with new taxes and regulations.[7] The gap between knowledge and power is filled with government rules and price controls.[8]

The findings of Henderson and Bain represented a revolution in economics. Economists have long ascribed such gains to economies of material scale and financial power. The logical response to that view is a massive assemblage of resources—land, labor, and capital—depicted in the poster art of socialist realism. But in the Bain model, volume does its work by increasing information and learning, knowledge and experience. Growth springs not from huge overcapitalization but from the responsiveness to customer demand that builds volume. Not only does volume create learning at the company, it also implies more interactions with more customers. The result is a process of customer learning that yields more uses of the product in a spiral of growing knowledge, power, entropy, and profit.

Economic growth springs not chiefly from incentives—carrots and sticks, rewards and punishments for workers and entrepreneurs.

The incentive theory of capitalism allows its critics to depict it as an inhumane scheme of clever manipulation of human needs and hungers scarcely superior to the more benign forms of slavery. Wealth actually springs from the expansion of information and learning, profits and creativity that enhance the human qualities of its beneficiaries as it enriches them. Workers' learning increasingly compensates for their labor, which imparts knowledge as it extracts work. Joining knowledge and power, capitalism focuses on the entropy of human minds and the benefits of freedom. Thus it is the most humane of all economic systems.

George Romney intuitively grasped these principles in his insistence that the American economy is made for man, not man for the economy. At Bain & Company, though, his son carried this insight to a higher level in his recognition that the essential role of the consumer, like that of the entrepreneur, is not merely spending but learning.

The Extent
of Learning

IN A LANDMARK speech to the Mont Pelerin Society marking the bicentennial of *The Wealth of Nations* in 1976, the Chicago economist George Stigler drilled in on what he regarded as one of Adam Smith's canonical tenets: the idea that it is "the extent of the market" that shapes and summons the economic "division of labor" and all its cascading innovations. This insight, said Stigler, is "one for which he is overwhelmingly famous.... [It is] cited as often as any passage in all economics."[1]

Celebrating the expansion of free markets and their fruits, this concept is a cornerstone of prevailing theory, respected by most economists of all schools of thought. Yet freedom's triumph comes at a serious cost. Smith's vision of the entrepreneur as a tool of the market rather than its creator constitutes the original sin of what we call "demand-side" economics.

Smith posited an imperious pre-existing market "pulling" in the entrepreneur who serves it. It sees his business as expanding to the limits of this market and then monopolizing it. If the extent of the market determines innovation, and if the market is occupied to its full extent, why should innovation continue? This problem drove Smith and scores of his followers to predict that growth would end in a "stationary state."

If innovation is bounded by a pre-existing market, what force can dislodge the entrepreneur from his stronghold, now defended by all the capital, systems, and workers that his market affords? The obvious conclusion is that only the regulatory state—with its anti-trust powers—can prevent the deterioration of a free economy into an oppressive regime of giant monopolies too big to either innovate or fail.

Perplexed by the near absence of such monopolies, economists from Smith's day to our own have conducted an endless search for near-monopolies, quasi-monopolies, oligopolies, and other elusive bastions of private power. They typically focus on gigantic companies—such as Standard Oil, U.S. Steel, General Motors, IBM, Microsoft, and now Google—just before their dominance collapses. All of the intricate but ultimately pointless solutions to the problem founder on Smith's crucial error—the demand-side premise itself.

The notion that markets define technology, that demand creates supply, is a profound fallacy that leads to endless mischief and meddling in entrepreneurial activity. Encountering that fallacy in traditional economics thirty years ago prompted me to write *Wealth and Poverty*.

The market creates neither the product nor the process of production. The entrepreneur and his product create the market. The priority of entrepreneurs to markets is suggested in the development of Western law, as Harold J. Berman observed in *Law and Revolution*: "The initial development of mercantile law was left largely ... to the merchants themselves, who organized international fairs and markets, formed mercantile courts, and established mercantile offices in the new urban communities that were springing up throughout Western Europe."[2]

The idea that entrepreneurs create markets, not the other way around, is not new economics. Good old-fashioned supply-and-demand curves would never meet, except in traumatic scarcities, if the size of the market—demand—did not vary with the process of production. Illustrating the point is Adam Smith's own famous description of a pin factory. He marvels at the huge gains in productivity of labor achieved through breaking down pin production into an array of specialties and expanding the scale of manufacture. Ten men organized in a factory can manufacture thousands of times more pins in an hour at a radically lower cost than can one man performing all the steps one at a time.

A pin that costs a penny is a wholly different product with a wholly different market from a pin that costs a tenth as much, even as a transistor that costs ten dollars is a different product with a different market from one that costs a ten-millionth of a dollar. The lower price extends its own market.

One of Bill Bain's first assignments at the Boston Consulting Group provides a good example of this phenomenon. In the mid-1960s at Texas Instruments in Dallas, Navy veteran Patrick Haggerty was a leading proponent of the learning curve for the production of electronic components. There had been no apparent market for any of TI's initial semiconductor products, such as germanium diodes and the first silicon transistors. These novel devices required redesign of systems that were then using advanced miniaturized vacuum tubes, which were well down their own learning curves. With Haggerty's support, Bain's team devised a strategy of aggressive pricing drastically below initial costs. Assuming that the cost of each device would plummet according to the learning-curve formula, TI surged to leadership in the industry. Its transistors came to epitomize the power of the curve and its ability to create new markets.

At about the same time, in 1965, Gordon Moore, the young director of R&D for a subsidiary of Fairchild Camera and Instrument, made an explosive prediction in an article for an industry journal. Using research from his young associate from Caltech, Carver Mead, Moore prophesied an annual doubling of the number of transistors that could be put on a single silicon device. He was not prompted in this prediction by any estimate of demand in any already available market. "Moore's Law," as Mead dubbed it, was later altered to predict a doubling of capacity every eighteen months and finally every two years. It could just as well have been called "Henderson's Law" or "Bain's Bonanza," and it all came on the supply side at a time when the market for transistors was almost nonexistent.[3]

All that differentiates Moore's Law from the Bain-Henderson theory is the length of the learning curve he was working with. Chips rode a faster learning curve than any other product. By contrast, it took not eighteen months but five years, beginning in 1915, for automobile production to double, and another five years to double again. But the story line is the same: the invention of the automobile and its falling price expanded the market for cars despite the absence of paved roads, gasoline stations, or maintenance garages.

In studying this process, economists usually stress the elasticity of demand (how much more of the product is purchased when the price drops). But incomparably more important is the elasticity of supply. This is determined by the entrepreneur's ingenuity, the availability of key resources, and the physical possibilities of the materials and systems.

When it comes to resources, as Moore was also the first to point out, integrated circuits enjoy a huge advantage over other products. They are made chiefly of silicon, oxygen, and aluminum, the three most common elements in the earth's crust. Unlike farmers or freeway contractors, who inevitably face diminishing returns as they use up soil and real estate, microchip manufacturers use up only chip designs, which are products of the human mind.

The magic of miniaturization and the learning curve continuously expand the market for chips. Take the case of Fairchild's 1211 transistor, which enabled television sets to be fitted with UHF tuners. In the early 1960s, each TV contained only one transistor. Potential sales were limited more or less to the number of households on the globe, so the market extended to mere billions of transistors. At that volume, discrete transistors like the 1211 could decrease in cost to the price of their packages, about a dime apiece, but no further.

With the development of the integrated circuit, however, you could put ever-expanding numbers of transistors together on a single silicon sliver. Today just one large-screen television set contains trillions of transistors, more than all the TVs in the world a few decades ago. The information advances Shannon made possible have put video capabilities in every smartphone and tablet, each with hundreds of billions or trillions of transistors, each one a thousandth of a penny or less in cost. The bold adventure on the supply side initiated by Moore, along with his Fairchild colleague Jerry Sanders, has climaxed in an information economy.

The Moore's Law pace of advance continued beyond the billion-transistor chip that Mead predicted. He thought that spontaneous quantum tunneling—where the silicon substrate itself begins erratically conducting electrons—would bring progress in densities to an end. Mead was correct about the physics. But in the modern economy, physics provides less room for advance than information theory does.

In this case Mead may have underestimated future contributions of information theory. Shannon had shown how to produce reliable systems with unreliable components by using sophisticated techniques of error-correction and redundancy and by lowering the level of voltages. Including millions of embedded redundant transistors and error-correction logic, modern memory chips now function at close to a single volt, compared with ten volts in the early days. Using less reliable transistors at lower power, with more redundancy and logical structures for built-in self-testing, the industry moved on to ten billion transistors and beyond. Now the end is near for flat or "planar" devices, and the industry is advancing into three-dimensional chips.

As physical advances become harder to achieve in manufacturing, productivity gains increasingly stem from applications of information theory. These take the form of new computer architectures, robotic efficiencies, and compression schemes. At the end of this process is new software virtualization, where advances in computing hardware enable the replacement of a once physical manufacturing process with a software abstraction. Memory, processing, and networks are virtualized. A library book or a compact disc or a DVD or a computer game or a manufacturing catalogue or a programming manual is coded and compressed and moved from its shelf into a computer memory, or it is transmitted across the net to a screen or another machine in the cloud or in an automobile. It is virtualized.

This same kind of virtualization is now transforming manufacturing. As Chris Anderson documents in his book *Makers: The New Industrial*

Revolution[4] and demonstrates in his company, 3D Robotics, virtualization of manufacturing proceeds apace. The development of "printers" that produce three-dimensional objects on the basis of digital instructions is increasingly making physical manufacture widespread in the same way that personal computers made the manufacture of informational objects widespread. It is this breakthrough—not a preexisting market—that allows manufacturers to adapt their products to the needs of each individual in the same way that printing intellectual works or graphics enabled an individual with a workstation to compete with publishers around the globe.

Entrepreneurial innovation extends the market to the boundaries of human activity. Just as every human being today could, theoretically, own a cheap printer, every human being in the coming era will be able to possess a manufacturing printer. In the climax of the industrial revolution, goods and services will be digitally virtualized and specialized for each individual user.

Capturing the explosive increase in efficiency resulting from such virtualization—mixing mind and matter, information and energy—is once again the experience curve. Governing both parts of the mixture of information and energy is entropy. Informational entropy measures the content of a message through the "news" or surprises it contains—the number of unexpected bits. The "division of labor" is measured by this entropy: the more specialized the production, the more the entropy and the market extend.

High-entropy inventions capture the *eureka* moments of enterprise—the light turning on in the mind that enables the next new thing. Surprising high-entropy breakthroughs win the economic glory. But for the guts of a full experience curve, you also need routinized production. While in communications you want unexpected news (high entropy), in a mass manufacturing process you want predictability (low entropy).

Moore's Law illustrates more dramatically than any other experience curve the interplay between high-entropy products and profits and low-entropy conduits and carriers. High-entropy entrepreneurial ideas and surprising new concepts in chip design cannot yield massively growing companies and profits without mastery of new production processes. Here in the factory or wafer fab, surprise usually means something bogging down or breaking. On the production line, surprises are often bad news.

Thermodynamic entropy measures wasted heat and movement—disordered and unrecoverable energy. This is the concern of moralistic movements for higher efficiency, recycling, and sustainability. But you cannot slow the processes of physical entropy merely by issuing high-minded restrictions on physical waste, suppressing emissions, and hobbling production with recycling rules. You cannot have a more efficient economy by suppressing entrepreneurial knowledge with government power. You have to permit and enable learning curves that reduce both forms of entropy at once, constantly multiplying and extending markets.

High informational entropy necessarily produces high physical entropy. At the outset of any fabrication process, uncertainty is high. No one knows how hard the machinery can be pushed; managers must supervise closely, keep large reserves of supplies on hand for emergencies, and maintain high manufacturing tolerances or margins for error. All this material inefficiency comes from uncertainty of information about the process. Without a substantial body of production statistics over time setting a low-entropy standard, managers are unable to distinguish a crippling defect, recurring in perhaps one of ten cases, from a trivial glitch occurring once in millions.

In any industrial experience curve, the two forms of entropy—energy waste and informational uncertainty—are being reduced. The combination of these two anti-entropic trends accounts for the 20- to 30-percent improvement in productivity. Radically reducing both forms of entropy in every conduit and channel of the economy, experience curve advances enable a vast and unprecedented rise of high-entropy content and communications, industrial inventions, and intellectual progress. All this occurs chiefly on the supply side and it is the source of expanding markets.

Enabling these advances of creative entropy are the reductions of physical entropy. As Moore's Law moves transistors closer together, wires between them become shorter. The shorter the wires, the purer the signal and the lower the resistance, capacitance, and heat per transistor. The result is a decline of physical entropy. As electron movements approach their mean free path—the distance they can travel without bouncing off the internal atomic structure of the silicon—they become faster, cheaper, and cooler until physical entropy approaches zero. The noise level plummets. Quantum tunneling electrons, the fastest of all, emit virtually no heat.

Thus, the very act of crossing from the macrocosm to the microcosm meant the creation of an industrial process that burst free of the bonds of thermodynamic entropy afflicting all other industries. In the quantum domain, as individual components became faster and more useful, they also ran cooler and used less power and lower entropy.

Understanding that Moore's Law is not a special case but a precursor—an experience curve applying to all human innovation—we can see that this unprecedented change is not a blip but a *beginning*. From processors to storage capacity, every technology touched by integrated electronics has advanced at a radically new speed. Today, in fact, the eighteen-month pace of Moore's Law appears slow compared with the often three-times-faster rate of the advance of optics.

Emerging as the spearhead of global progress in market extension through more effective communications is the fiber-optic technology called wavelength division multiplexing. WDM combines many different "colors" of light, each bearing billions of bits per second, on a single fiber the width of a human hair. The best measure of the technology's advance is lambda-bit kilometers, multiplying the number of wavelengths (lambdas) by the data capacity of each and the distance each can travel without slow and costly electronic regeneration of the signal.

In 1995, the state of the art was a system with four lambdas, each carrying 622 million bits per second some three hundred kilometers. In 2002, a company named Corvis introduced a 280-lambda system, with each lambda bearing ten billion bits per second over a distance of three thousand kilometers. This was an eleven-thousand-fold advance in six years. In 2012, systems were launched with each wavelength carrying 100 billion bits per second. With several hundred fibers now sheathed in a single cable, a fiber installation can now carry more than a month's worth of 2002's Internet traffic in a single *second*.

While the power of microelectronics spreads intelligence through machines, the power of communications diffuses intelligence through networks—and not just computer networks but companies and societies—once again expanding markets across the global economy. And unlike silicon transistors, with their mass and expanse, photons are essentially without mass, completing the dematerialization that began with semiconductors. Physical entropy collapses, unleashing tides of high-entropy communications and ubiquitous markets.

The ultimate low-entropy channel, photonic carriers include the broad range of wireless signals. All can multiply without weight in the same physical space. They are largely non-rival physical functions that are not consumed as they are used. Virtually any number of colors can eventually occupy the same fiber core or wireless cell. The new magic of optics feeds on the ultimate convergence of rock-bottom physical and information entropy, perfect sine waves of electromagnetism that enable the economy to plunge down curves of experience without mass or resistance through worldwide webs of glass, light, and air.

By virtually eliminating physical entropy in computers and communications, this world of bandwidth abundance shifts the entire focus of innovation from physics to information. Entrepreneurs can magnify and accelerate the pace of high-entropy creation to levels unprecedented in human history. Markets expand to universality. Bain's analysis means that tax-rate reductions and other measures enlarging the opportunities of enterprise can move an entire economy toward a far higher level of learning and informational efficiency.

Every company can move further down its curve. Every worker can enhance his performance through increasing his units of experience. As engineers saturate the possibilities of physical theory, they move down the curves of learning in information theory. In their efforts, they create the market and extend it until the product is mature.

In this continuing process, the only limits come on the supply side. Physical entropy reaches its limit when all energy levels are equal and no further power can be extracted. Information entropy reaches its limit when all the bits are equally likely and no further knowledge can be captured.

These limits of knowledge and power apply to every enterprise. They represent the ultimate extent of the market that can be launched down a particular curve of learning and experience. When they are reached, a fresh leap of intuition and experiment down a new learning curve is needed.

From Adam Smith's pin factory to Moore's Law of microchips, the division of labor drives the extension of the market, not the other way around. Supply creates its own demand through the proliferation of goods and services down the curves of learning, entropy, and imagination.

The Light Dawns

THE CENTRAL SCANDAL of traditional economics has long been its inability to explain the scale of per capita economic growth over the last several centuries. It is no small thing. The sevenfold rise in world population since 1800 should have attenuated growth per capita. Yet the conventional gauges of per capita income soared some seventeen-fold, meaning a 119-fold absolute increase in output in 212 years. And this is only the beginning of the story.

The leading economic growth model, devised by the Nobel laureate Robert Solow of MIT, assigned as much as 80 percent of this advance to a "residual"—a factor left over after accounting for the factors of production in the ken of economists: labor, capital, and natural resources.[1] In other words, economists can pretend to explain only 20 percent of the apparent 119-fold expansion.

Apparently depressed by the shrinking domain of their discipline, many economists have turned to impugning the growth. Some charge that industrialization reduces workers to misery; others assert that the benefits of growth are outweighed by environmental damage. From Pastor Malthus to the Club of Rome's *Limits to Growth*; from hysteria over DDT, PCBs, and natural gas "fracking"; to continuing bouts of chemophobia and population panic; the achievements of capitalism have suffered a long series of detractions. The factitious and febrile campaign against global warming is only the latest binge of self-abuse among the children of prosperity.

When these dismal themes have failed to satisfy, leading economists from Smith to Samuelson and from Ricardo to Galbraith have portrayed growth as an aberration on the path toward a stationary state, when progress ends. They are still at it. Robert J. Gordon, an eminent growth theorist from Northwestern University, recently electrified his audience with a major paper published by the National Bureau of Economic Research, declaring that growth was an episode unique in human history that has at last come to an end.[2] He discerns our doom in "six headwinds"—waning industrial and information revolutions, and the nemesis of "climate change." This conclusion would seem to extenuate the failure to explain growth, except as a curious and temporary anomaly in human affairs.

In *Knowledge and the Wealth of Nations* (2006), David Warsh, the eminent economic journalist, boldly rejects this perennial thesis.[3] He reports on a contrary campaign of economists to offer an account of economic advance as the effect of knowledge, which, unlike land, labor, and resources, faces no obvious limits to growth. He presents evidence that capitalist growth has in fact been a *thousand times* greater than is registered in the conventional data.

This point of enlightenment has been reached, however, only after an arduous climb. After the luminous achievement of Adam Smith and the original *Wealth of Nations* in 1776, reports Warsh, academic economists pursued a benighted Lilliputian adventure, with many digressions, that blinded them to the immensity of the phenomenon they must explain and to the power of entrepreneurial creativity and knowledge.

As with Audrey Hepburn's blind character stalked by a killer in *Wait Until Dark*, we wonder avidly what obvious object—inventive breakthroughs, creativity, or learning—they will bump into next without recognizing it? Down what new flight of stairs—Marxism, convex supply curves, or environmental exhaustion—will they tumble in search of enlightenment? What dismal calculations of diminishing returns will they pursue in the basement?

All is not lost, however. To break the suspense, Warsh introduces us to the economist William Nordhaus of Yale. Nordhaus finally finds the switch, and the light floods in. In a paper titled "Do Real Income and Real Wage Measures Capture Reality? The History of Lighting Suggests Not,"[4] he shows that for half a million years, from cavemen's fires to the candles that illuminated the palace of Versailles, the labor cost of a lumen-hour of light dropped by perhaps 75 percent. Then between 1711 and 1750, the British government embarked on an anti-light program, taxing candles and windows, increasing the cost of illumination up by 30 percent. On the eve of the Industrial Revolution—the very time that Malthus and Ricardo were formulating their lugubrious theories—Great Britain entered a "little dark age," a period of stagnation when no one envisioned the possibility of major economic advances.

How could a tax produce a dark age? Traditional economics has an answer, and that answer is not wrong: tax a thing and you get less of it; tax light and darkness encroaches. It is hard-earned wisdom, often forgotten today even by sophisticated thinkers like David Warsh. The experience curve might deepen the explanation: slowing the volume of candle sales by a tax also retards candle innovation and price reduction.

What followed, however, renders these sensible observations nearly trivial. The little dark age was not dispersed by repealing the tax and restoring sensible incentives to candle makers, prudent though such a policy would have been.

What happened was that everything changed. Over the course of the nineteenth century, Britain became increasingly open to innovation and trade around the globe. The labor cost of light plunged, with gas light costing one-tenth as much as candlelight and kerosene light one-tenth as much as gas light. Fossil fuels were the salvation of the whales. The arrival of electricity in the 1880s produced another thousand-fold drop. In other words, one of the most astonishing increases in wealth in the history of mankind, a million-fold increase in the abundance and affordability of light itself, took place through a process of invention and innovation not comprehensible to economics in the ordinary sense of that word. No mapping of economic efficiencies can explain it. Economics could give an account of the sudden abundance of light only if it focused on the conditions of human creativity itself.

Why isn't economics "about" creativity and entrepreneurship and the shattering of equilibrium by sudden, radical explosions of new information? One reason is that economists missed the revolution itself. Relying on defective price measurements that failed to gauge the cost-effectiveness of the new technologies of illumination, they remained in the dark. They groused about the rise of the British national debt, which reached 250 percent of GDP in 1820, and decried the immiseration of workers in those "dark satanic mills." They concentrated on money prices rather than real labor costs—how many hours workers had to labor to buy light. Dwelling on liabilities rather than assets, the economists almost completely missed the fabulous experience curve in the data of growth. Over the course of the nineteenth century, as Nordhaus showed, the cost of light plummeted to merely one-tenth of 1 percent of its level in 1800. It was a precursor to the famous Moore's Law of advance in microchips with its doubling of computer cost-effectiveness every eighteen months.

Warsh exclaims, "The traditional story was off by *three orders of magnitude*, or a factor of a thousand-fold!"[5] How did it happen? One perhaps cynical explanation—not entertained by Warsh—is that denial of the centrifugal feats of entrepreneurial knowledge, distributed unpredictably across every free economy, protects the centripetal power of kings, bureaucracies, politicians, and other purchasers of economic influence. Demand-side GDP data tend to miss all the most important technological revolutions and thus foster zero-sum thinking oriented toward government beneficence and redistribution.

As Norhaus concluded, economists tend to be blind to the blazing curves of learning that abound in the economy. Technological revolutions are absent from price indices. Nordhaus recognized that virtually all of economic growth is driven by entrepreneurial and scientific creativity.

Nordhaus wrote in 1993. Perhaps over the last twenty years the economic prophesies of limits to growth and innovation have finally come true. Robert Gordon declares that innovations in information technology have far less impact on human life than previous breakthroughs like kerosene lighting. But information tools reach far beyond the computing and networking equipment that are their most visible embodiments to embrace the fundamental processes of economic growth.

In information theory, new shoots sprouting from previous learning produce *conditional entropy*. The move from whale oil to kerosene, for example, entailed creation of an almost entirely new system of illumination. The conditional entropy represented a degree of surprisal that was contingent on previous learning and embodied in standards and infrastructure, from the docks at Bedford, Massachusetts, to the oil rigs at Spindletop. Both forms of fuel, though, relied on the refining processes and lamp ignition chemistries of previous technologies. Learning curves compound. Both physical and informational entropy diminish with each new launch down a curve.

Think of the caveman at the beginning of Nordhaus's tale. Uncertain of the weather and his needs, he is gathering and breaking up, lugging in, and storing piles of wood in his cave in order to have a little light on his dinner at night. He loses energy as physical entropy at every phase. His utter lack of certainty (his high-entropy informational life among saber-toothed tigers and other predators) leads to an enormous waste of food, fuel, heat, and light.

Thousands of learning curves later, each reducing uncertainty in some facet of light production, a woman walks into a room. Light-emitting diodes (LEDs) immediately illume the space at a physical entropy loss millions of times less than that incurred by the caveman's fire. The pin factory is now manufacturing these LEDs, and the process of learning at each step of fabrication removes uncertainty and energy loss, informational entropy and physical entropy.

The next step in the story of lighting is already upon us in early 2013 with the move of Moore's Law and silicon-based technology into the lighting industry in the form of *silicon* light-emitting diodes. As Eric

Savitz of *Forbes* and Bill Watkins of Bridgelux report, "Lighting fixtures, freed from fragile and bulky bulbs, will see a revolution in design. You will see stairs and cabinets that come with embedded thin and rugged lights. Lighting in your home will be programed [*sic*] to emit a range of colors so you can change the mood whenever you feel like it."[6] Networked LEDs can be combined with silicon transistors to "detect motion, predict maintenance or figure out how many people are in a room." In the familiar pattern of Moore's Law, lights will get smaller, faster, cheaper, and more functional.

Commercial production of white-light LEDs began in the 1990s, and billions of LEDs have been produced on sapphire substrates coated with a thin layer of a crystalline material called gallium nitride, which converts electrons to photons—electricity to light. The cost of a kilolumen (1,000 lumens), Watkins reports, was $302 in 2000. By 2011, the price had dropped to seven dollars. Measured in lumens per watt, the efficiency of the technology rose from fifteen in 2000 to 160 in 2011. But building light bulbs on expensive sapphire wafers limits use to relatively high-end applications.

"Enter gallium-nitride-on-Silicon," proclaims Watkins. GaN-on-Silicon is a method of growing gallium nitride on conventional silicon wafers. Watkins's company, BridgeLux, devised a way. Toshiba and other companies are studying similar concepts. By switching from sapphire to silicon wafers, the wafer size jumps from two or four inches to eight inches, expanding the surface area of the wafer from 12.56 square inches to over fifty square inches. Since processing a two-inch wafer takes roughly as long as processing an eight-inch wafer, bigger wafers enable expansion of production in the same facility by nearly thirty-six-fold, drastically increasing capacity while reducing unit costs on the pattern dramatized by Nordhaus's historical saga of the plummeting cost of light.

All these phenomena of learning, entropy, and growth, however, failed to penetrate the world of conventional economics. Turning to ever more abstruse mathematics in the twentieth century did not solve the problem. As Paul Krugman observed, "Economics inevitably and understandably follows the line of least mathematical resistance"—looking for wallets under the most luminous lampposts and leaving whole continents, such as Africa and entrepreneurs, in darkness. During much of the century, they clustered in tribes around three or four major totemic light sources: Adam Smith, with his magical self-extending markets; John Maynard Keynes, with his amazing self-fulfilling demand; Kenneth Arrow and his disciples, with their Keynesian growth models of mostly unchanging products. Meanwhile Hayek and Samuelson defended the spontaneous order and equilibrium of Walras and Marshall.

Understanding the errors may afford new light.

Keynes Eclipses Information

"AS AN ECONOMIC HISTORIAN who has been studying American capitalism for 35 years, I'm going to let you in on the best-kept secret of the last century," James Livingston wrote in a May 2011 column for the *New York Times*. The secret was this: "Private investment—that is, using business profits to increase productivity and output—doesn't actually drive economic growth. Consumer debt and government spending do."[1]

It's a venerable idea, elegantly expressed by Keynes's contemporary Michal Kalecki in his famous profits equation. The communist son of a failed Polish businessman, Kalecki based his insight on permutations of the tautology that income equals spending. "Your income is someone else's spending" is a cardinal truth of Keynesianism. Such eminent figures as Paul Krugman believe it is a fair summation of economic reality. It describes a circular economy in which the outputs of some become inputs for others. Breaking down income into profits (for capitalists) and wages (for workers) and apportioning spending into investment and consumption, Kalecki shows that profits must equal investment minus savings plus dividends. Thus *government* savings (surpluses) subtract from profits, while government "dissaving," as economists call it (deficits), adds to profits.

Here's where Livingston comes in. Although he abhors profits, he is interested in growth, which accompanies profits. Since government spending and dissaving, represented by accumulated deficits, have increased more than any other form of dissaving since the 1930s, all profits and growth can be attributed to them. It's all in the equation—deficits produce the growth of GDP; spending causes income.

To see how this process plays out, we can consider an explanation of the economy that Krugman lays out in two books, *The Return of Depression Economics* and *Stop This Depression Now!*[2] He makes the authoritative argument that a depression—like the one that began in 2008—can always be halted by a sufficient increase in government spending. His book on the return of depression economics was undeniably prescient. Krugman plausibly presents himself as a new Keynes, offering bold remedies to a timorous financial establishment.

For all the partisan rancor in his *New York Times* columns, Krugman is an exemplary economist and a lucid writer, so the professional frailty revealed in these books is noteworthy. He sums up the dominant theory, in terms that many conservatives would readily accept, in his factual parable of "the Capitol Hill Babysitting Cooperative." It is, he frankly states, his "favorite economic story," bearing the most profound and important practical truths. It illustrates the governing paradigm of his work. "It is life changing." Those who reject its lessons, he declares, are "utterly wrong."

The co-op began in the late 1950s with some twenty families joining to exchange babysitting services. During the 1970s it reached its peak membership of two hundred households, mostly congressional staffers. Rising tenfold in twenty years, its GDP achieved a compound rate of growth of roughly 9 percent a year. On entering the co-op, members received twenty coupons or scrips—repayable on leaving the group—each worth one half-hour of babysitting from other members. Since they had to return as many coupons as they received, they had to perform as much babysitting as they received. They also had to contribute fourteen hours a year of administrative work.

As Krugman describes it, the scrip gave them liquidity, the economic word for freedom. The members of the co-op could go out any evening they wished in the likely assurance that among one hundred and fifty or two hundred other participants someone would be willing to accept their scrip and tend their children.

For about twenty years, the venture was a success, solving the babysitting problem of these young Washington couples, who were perhaps gratified to know that their sitters, like ambassadors or Supreme Court justices, had been vetted by U.S. Congressional offices or committees. But as time passed, members showed an unexpected propensity to accumulate scrips—"saving them up in their desk drawers," as Krugman puts it[3]—perhaps to enable themselves to go out several nights in a row, and perhaps to avoid being trapped in the co-op as indentured sitters while they tried to pay back their quota.

In any case, these precautions illustrate what Keynesian economists felicitously call the "paradox of thrift." During an economic crisis, one person can increase his savings by forgoing consumption, curtailing his spending and paying down debt in a process called deleveraging. But when many people try to pay down debt or increase their savings at the same time, overall savings in the economy will decline because a general drop in spending causes a drop in incomes (and savings are the residual of incomes minus consumption). The paradox is that the more people try to save, the less savings there are overall.

In the same way, when too many couples sought to provide services—collecting scrip without spending their own scrip—the co-op suffered what Krugman describes as a depression, just like the depression that has bedeviled the United States since 2008.

From the parable of the babysitting co-op, Krugman draws the lesson that your income is someone else's spending, and your spending is someone else's income. If everyone is trying to save up scrip by babysitting but failing to spend scrip on babysitters, the babysitting market has to cut back its production to the diminished level of spending. There is no compensating rise in spending by other members caused by your own agoraphobic or thrifty refusal to go out. Your proffered supply of baby-sitting services does not generate demand. It doesn't induce increased investment in restaurants and theaters. According to Krugman, the co-op disproves Say's Law, the rule of classical economics that supply creates its own demand. As in the paradox of thrift, what *one* can do, *many* cannot, because one couple's income in scrip depends on another couple's spending its scrip.

Most of the coupons remain in desk drawers. Activity collapses. A large gap opens between the group's potential and actual output. Mass unemployment and massive unused capacity banefully coexist. It's a depression at the babysitting co-op.

Luckily, as Krugman reports, among the co-op's members were several economists! They got together and persuaded the co-op's management to increase the number of coupons—the co-op's "money supply"—by 50 percent. Each participant would receive ten additional coupons from the outset. This move satisfied the participants' nasty urge to save, with the result that they were willing to spend more freely. They were still obliged to return twenty coupons to the management on leaving the co-op, and their scrip income could still not exceed their scrip spending, but the participants were more liquid. Satisfied with their savings, they were ready to binge in a booming babysitting economy.

To Krugman, this fable conveys the most important economic wisdom: *It's all about demand.* Everything depends on the urge to spend. It is the "magneto," as Keynes dubbed it, or the "battery," as Krugman prefers, that starts the economic engine and gets the machine on the road. Just as a thirty-thousand-dollar car can be put out of commission by the depletion of a hundred-dollar battery, so a fifteen-trillion-dollar economy can be plunged into depression by "magneto trouble"—a downdraft in spending that results from a collective effort to save or deleverage. Just inject some more spending, like so much electric power, and all will be well. It does not matter who spends or how they spend or what they spend on. As Keynes wryly put it, the government can hire people to dig holes and fill them up again. It doesn't matter. The key is the power to spend.

It would be unfair to Krugman to imply that he does not grasp the limitations of a time-valued coupon that precludes changes in the price of sitter-hours. In the real world of money, his co-op would adapt to changes in the propensity to save by constant price changes adjusting between the supply and demand for sitters. To overcome this objection, he posits a Keynesian "liquidity trap," like interest rates of zero that cannot be reduced and that prevent the Federal Reserve from addressing the depression merely by printing money. No one will lend money at a rate of zero. The co-op script with its set price somehow constitutes a liquidity trap. It does fit Krugman's model.

The babysitting economy is a static monoculture of unchanging providers and consumers of services. It is based on the model of the circular flow of homogeneous media that somehow has captivated the most prestigious and sophisticated of economists. And, to quote Krugman, it is "utterly wrong." Let us return to Kalecki's and Livingston's version of Krugman's argument that demand is everything, with the corollary that government dissaving is the salvation of failing economies.

A tempting tautology for every sophomore economics student enamored with the promises of heroic government, and irresistible to liberal politicians and economists, the Kalecki-Krugman principle pervades much economic analysis. Every quarter, when the government releases its latest GDP figures, we hear the familiar refrain: "If the consumer stops spending and starts saving, we will be in big trouble—consumer spending accounts for 70 percent of the economy." The other 30 percent of what these economists call aggregate demand is attributed to government spending (23 percent), investment spending (10 percent), and net exports (-3 percent, as imports exceed exports). Because everything, in this view, depends on the eagerness of the demander to demand and the consumer to consume, Krugman calls almost weekly for more government spending, more government deficits, more monetary easing, and more debt.

In his *New York Times* column, Livingston—who would soon publish a book titled *Against Thrift*[4]—put dramatic numbers on this idea that demanders, and not suppliers, are the drivers of the economy:

Between 1900 and 2000, real gross domestic product per capita (the output of goods and services per person) grew more than 600 percent. Meanwhile, net business investment *declined* 70 percent as a share of GDP. What's more, in 1900, almost all investment came from the private sector—from companies, not from government—whereas in 2000, most investment was either from government spending (out of tax revenues) or "residential investment," which means consumer spending on housing, rather than business expenditure on plants, equipment and labor.[5]

Business investment is, meanwhile, an actual hazard:

So corporate profits do not drive economic growth—they're just restless sums of surplus capital, ready to flood speculative markets at home and abroad. In the 1920s, they inflated the stock market bubble, and then caused the Great Crash. Since the Reagan revolution, these superfluous profits have fed corporate mergers and takeovers, driven the dot-com craze, financed the "shadow banking" system of hedge funds and securitized investment vehicles, fueled monetary meltdowns in every hemisphere and inflated the housing bubble....

Consumer spending is not only the key to economic recovery in the short term; it's also necessary for balanced growth in the long term.... [W]e consumers need to save less and spend more in the name of a better future.[6]

This might seem like a crude form of Keynesianism, but it does not go far beyond Keynes's own views in expounding his paradox of thrift. Already a legendary figure in Britain for his swashbuckling leadership at the wartime treasury and as a member of the Bloomsbury Group, he wrote that although the amount of one man's saving "is unlikely to have any significant influence on his own income, the reactions of the amount of his consumption on the incomes of others makes it impossible for all individuals simultaneously to save any given sums. Every such attempt to save more by reducing consumption will so affect incomes that the attempt necessarily defeats itself." (Keynes himself harked back to several precedents for this idea, including an early-eighteenth-century poem, *The Fable of the Bees*—pithily subtitled *Private Vices, Public Benefits.*)

Looking for disequilibria to reconcile into equilibrium, disorder to transform into order, economists suppress the multifarious complexities of supply to create a homogeneous Keynesian spending model. But they're wrong to do that. As the economist and investor Mark Skousen (a proud descendant of Ben Franklin, apostle of thrift) has been documenting for decades, consumer spending is nowhere near 70 percent of the real spending in the economy. It is 70 percent of the putative spending on "final products," most of which economists define as "consumer goods." But consumer goods represent only the present flow in an economy that entails many stages of production over time, which Skousen inventively models.[7]

What Livingston sees as a decline in capital investment as a proportion of GDP is actually a huge rise in the productivity of capital, as each dollar of investment yields a greater increment of growth. After all, capital is not merely a flow of spending. It is a complex structure of productive inputs. The more cheaply the investor can finance his business, the better off his business will be, and the more productive and prosperous the economy. In the air transport and steel industries, for instance, it turns out that computers produce more growth with less expenditure than do more airplanes or steel mills. The increase in the productivity of capital reflects its ongoing renewal through information and innovation.

Skousen calculates total spending (sales or receipts) in the economy at all stages to be more than double the official GDP figure. By this measure—which Skousen calls gross domestic expenditures (GDE)—consumption represents only about 30 percent of the economy, while business investment (including intermediate output) represents over 50 percent.[8]

Retail sales, notes Skousen, are not even one of the leading economic indicators that the Conference Board tabulates each month: manufacturers' new orders, building permits, unemployment claims, average weekly manufacturing hours, real money supply, stock prices, the yield curve, new orders for nondefense capital goods, vendor performance, and index of consumer expectations. All of these indicators—even the index of consumer expectations—reflect not consumption but the early stages of production and business activity.

The compilers of the Index of Consumer Confidence ask consumers about business conditions and investment plans, not about their short-term consumption: Are jobs currently plentiful, not so plentiful, or hard to get? Within the next six months, do you plan to buy a new/used automobile/home/major appliance (all consumer durables with some investment characteristics)? Are you planning a U.S. or foreign vacation? Conspicuously absent are any questions about food, clothing, entertainment, and other short-term buying, which change little from month to month.

Creative business and informed investment drive the economy and the stock market. For cues to the future, ignore the consumer (except in his more important role as a producer), and analyze manufacturing plans, corporate profits, productivity gains, and venture capital.

The preoccupation with the consumer is merely a reflection of what Keynes identified as the enslavement of practical men to "some defunct economist"—in this case Keynes himself and his ideas about the futility of savings. But a recent study by the Federal Reserve Bank of St. Louis concluded that even in the short run, "a higher saving rate in the current quarter is associated with faster (not slower) economic growth in the current and next few quarters."[9] Savings reflect the profitability—the upside entropy—in the economy. Their growth is usually the result of unexpected cash flows, whose rise normally signifies prosperity. Willingness to save indicates consumers' preference for products to be supplied in the future over current goods and services. Satisfied today, consumers supply resources for consumption tomorrow. Savings both endow investment and defer demand. Thus it is not implausible for the St. Louis Fed to conclude, "[T]he growth rate of real GDP has been higher on average when the personal saving rate is rising than when it is falling."[10]

In response to Livingston's attack on thrift in the *New York Times*, Skousen wrote,

Which drives the economy, consumer spending or savings/ investment? Country A saw [nominal] consumption as a percentage of GDP rise from 60 percent to 75 percent since 1980, and the personal savings rate drop from 12 percent to 4 percent. Country B saw consumption as a percentage of GDP fall from 52 percent to 36 percent since 1980, and the personal savings rate rise from 34 percent to 53 percent. Country A is the United States, and Country B is China. I wonder which one grew faster in the past 30 years?[11]

Skousen chose to focus on China and the U.S. A Keynesian might respond that the two countries were in effect a single economic entity, with China benefiting from U.S. consumption. In his compendious texts authoritatively addressing the subject, Skousen cites many other examples of un-paradoxical thrift. Illustrating the point as well as China are the experiences of Singapore, Germany, Hong Kong, Switzerland, Taiwan, South Korea, and Brazil. The fastest growing economies in the world have normally been heavy savers.

Saving powerfully diverts consumption preferences from immediate goods to the array of intermediates funded by savings. Savings prepare the economy for a long future of growth, compensating for the dwindling harvests of consumption in a world of impetuous spending. Even during a depression, as Skousen points out, "the old time virtues of retrenchment, getting out of consumer debt, selling off assets, cutting costs, and increasing savings can allow the nation to start back on the road to recovery.... [A consumer] spending spree in the depths of the depression would only make matters worse by reducing the amount of funds flowing to the more depressed capital sector and thereby bankrupting people."[12]

Skousen's long crusade against the paradox of thrift has been vindicated in scores of countries over many decades. But in two recent books on the crisis of capitalism, my friend Richard Posner, an inspiration for my emphasis on the altruism of enterprise in *Wealth and Poverty*, failed to read him and joined Krugman in sharply condemning supply-side economics. Following Keynes, both Posner and Krugman drilled into the very foundations of supply-side theory, Say's Law. Although they imagined this foundation is permeable, they both blunted their drills on its granitic truth.

Keynes memorably rendered this principle (named for the embattled French Revolution-era economist and entrepreneur Jean-Baptiste Say) as "supply creates its own demand." Then he ambiguously refuted it. Half a century later, Posner, Krugman, and all the other Keynesians still feel the itch to refute it again and again. Posner grandly proclaims that the depression of 2008 finally disproves Say. It's about time. Like Krugman, Posner assumes that Say's Law, playing out across an entire economy, means that there is always enough demand to purchase any given level of supply—that a depression is impossible.

"In a stark refutation of Say and vindication of Keynes," says Posner, people are saving more but government borrowing overwhelms their savings. The result is that aggregate savings—public plus private—are negative. "Negative savings, negative private investment, an incredible [though slightly declining] ratio of household debt to disposable income (1.25 to 1), massive government borrowing to finance private consumption—these are signs that we are propping up consumers for yet another binge."[13]

Posner and Krugman are confidently refuting a model of Say's Law based on a notion of a circular flow. But Say was a follower of the French "physiocrats," whom he named, and their idea of a circular flow had nothing to do with the mere flow of funds. Deeply grounded in microeconomics, Say's theory assumed an economy in which physical goods are exchanged for other physical goods behind a veil of monetary transactions. "Products are paid for with products," as he put it.

Say knew that at times an economy, often under political pressure, would devote too much of its resources to the production of goods that turn out to be unneeded (subsidized housing, windmills, sugar, and ethanol, to cite some current examples), and too little to needed items, such as oil, gas, and nuclear power. As Say well understood, such misallocations would lead to recession. But always for Say, the basis of economics was the production of goods: "The encouragement of mere consumption is no benefit to commerce; *for the difficulty lies in supplying the means, not in stimulating the desire, of consumption* [emphasis added]."[14]

Posner and Krugman take these microeconomic truths and transform them into the macroeconomic figment of an economy consisting of circular flows, where every output is the input for a subsequent phase in the process. All macroeconomic circular flow arguments assume two propositions:

1. Inputs, which are former outputs in the chains of production and consumption, retain their stated value. (The purified silicon ingots and engineering skills lavished on Solyndra solar cells, for example, will be worth as much when they are converted to the output of Solyndra's cylindrical solar systems).
2. Input value is independent of the inputter and his skill, that is, the information and knowledge embodied in the product. In other words, capital is mere spending, independent of capitalists and their specific creative forms of capital goods. The Keynesian assumption is that savings and investment constitute a flow of spending power rather than a complex and specific fusion of knowledge and news, data and design, achieved by entrepreneurs.
3. The value of government spending, measured by its cost, is comparable to the value of private spending, which reflects prices people were voluntarily willing to pay.[15]

The circular flow is a communications system with an unlimited channel—lacking creators, noise, memory, redundancy, and information. Such a communications system will necessarily fail if the message is diverted at a node on the line. On these assumptions, Krugman and Posner, following Keynes, are correct to worry that some saved output will leak out because it will not be reinvested as input or used for consumption. It will be diverted into savings or wasted.

But the circular flow is an economic metaphor that ignores all the specific information and knowledge that make economies work. Capitalists, investors, and managers vary widely in their skills and command of information. The channel is inevitably noisy. Constantly depleting the circular flow are waste, friction, mistakes, losses, depreciation, capital excess—the new worldwide headquarters or overbuilt factory or corpo-rate solar bauble or proprietary enterprise network or plush supercom-puter facility that signals an executive's indulgence, launched in the absence of profitable ideas or in the presence of undue taxation of profits.

Inputs suffer noise, distraction, and degradation at every point. Yet these obstacles do not stop growth. While many managers, often respond-ing to governmental subsidies, are wasting their resources, creative entre-preneurs in garages and at kitchen tables are concocting surprises in which capital is multiplied. For all the billions of dollars wasted on Solyndras and windmills, a few score million were invested in natural gas fracking experiments that yielded trillions of new cubic feet and hundreds of billions of dollars' worth of new asset value from Pennsylvania to North Dakota.

And then there was Gregg Robertson in South Texas, whose story was reported in *Forbes*. He noticed dust on the floor of a sprawling warehouse the size of three football fields. In 1952, five years before Robertson was born, a Phillips Petroleum drill bit brought to the surface sediment depos-ited on the sea floor more than sixty-six million years ago. Robertson had firsthand knowledge of two generations of South Texas oil field ventures. In 2008, he sent a sample of shale fragments from the Phillips well in La Salle County for analysis and set off the boom that hit fever pitch in 2012, yielding possibly the best "tight oil" play in the United States. Tight oil is the type of light crude distilled from shale formations, once regarded as irretrievable but now transforming sleepy villages into boomtowns, turn-ing penurious ranchers into millionaires, and opening new opportunities for thousands of people across South Texas.

"People are using it to understand the potential around the world," said Steve Trammel, an expert on unconventional oil and gas exploration for the business information and analytics provider IHS. "There could be as much as 500 billion barrels of recoverable fuel in this tight oil play."[16]

Efficient users of capital like Gregg Robertson are compensating for the waste, the distortion, and the destruction of value caused by the spu-rious "investment" directed by Washington, which becomes inadvertent consumption. Most innovative entrepreneurs stand outside the Krugman-Posner circle games. Nearly all transformative figures in business are only thinly capitalized in the conventional sense. The entrepreneur is the savior of the system because he capitalizes himself. He is his own most important capital. What we call intellectual capital is also informational and moral capital: diligence, good judgment, imaginative appreciation of others' inputs and needs.

Socialists believe their mission is to seize capital for the masses. But the great secret of capitalism is that, detached from a capitalist, there is no capital. To create wealth, knowledge and power must be merged. This is true even of great fortunes and great companies replete with conventional "capital" assets, because those assets wither and die swiftly without capitalists to tend them.

The unity of inputs and inputters is more apparent in the case of the entrepreneur than of the financial capitalist. Entrepreneurial profit, entropic profit, is not simply the transformation of input into output; it is the surprising excess of output over input. But this excess itself is in a sense an illusion. Measured profit is the excess of output over *accounted* inputs. The very nature and role of the entrepreneur is to infuse the venture with inputs—informational capital—that never make it to the accounting statement, and that all too often cannot be measured until the originator is retired or dead and a Walter Isaacson comes along to examine the scenes of creation.

This entrepreneurial capital—the unaccountable input of the innovator—is abundant in free economies and makes up for all the inevitable leakage out of the circular flow. Indeed, this entrepreneurial information and energy is the only source of restoration that prevents the circular flow from running down into inanition.

When the circular flow seems healthy, it is only because we do not notice that it is being constantly replenished. Such constant replenishment and revitalization by new information and knowledge is the only solution to the dissipation and physical entropy that is normal at all times but has become a crisis in our own.

Fallacies of Entropy and Order

IS THE SECOND law of thermodynamics—the one about physical entropy—true? Do all things tend to disorder? Is the universe in a steady state of decline? Is it moving step by step to randomness? Are form and structure steadily stumbling down the stairway of form into the chaos of a wispy gas?

Virtually everyone in the world of conventional science and its media satellites thinks so. The second law is perhaps the most widely accepted proposition in all of physics and chemistry. It imposes a time frame on a set of physical laws and equations that are otherwise time-neutral. It is self-evidently true for breakfast at your favored café, from the sugar and creamer suffusing your coffee to the eggs in your burrito. Extended into information theory, the second law supplies the mathematical formula that is the key to Shannon's work, and it is at the heart of this book.

Regardless of its unquestioned scientific utility, however, the second law is philosophically pernicious. It stems from a futilitarian zero-sum view of the universe. It supplies a central theme for an environmentalism that assumes the decline and exhaustion of all natural resources. It reduces credulous biologists and neuroscientists to the intellectual penury of a materialist superstition that denies the objective significance of their own scientific thinking. It leads unwary physicists into vain detours and aporias, with no empirical grounding, such as the so-called "multiverse," an infinite landscape of multiple parallel universes.

This venerable law of physical entropy, the lordly second law of thermodynamics, is not only futile and demoralizing; it is ultimately wrong. Entropy and disorder in the universe do not evidently tend toward a maximum.

So audaciously argues one Howard Bloom, a prolific author, inveterate atheist, lifelong Einstein scholar, amazing polymath, capitalist carouser, and grand master of mathematical logic and physical learning. "In fact," declares Bloom, dismantling the entropy law in his 2012 book *The God Problem*, "the very opposite is true. The universe is steadily climbing up. It is steadily becoming more form filled and more structure rich."[1]

From an imaginative slouch at a cosmic café table in Brooklyn where he sits ogling the passage of undulating particles at the beginning of the universe, Bloom confesses,

> When protons and neutrons became the new holy trinities of the material world—elementary particles [hydrogen, helium, and lithium]—I, the down-to-earth grump, was stunned.... When you've got a mess of particles slamming, banging, and bouncing, you are going to run into the second law.... You are going to end up with a random soup.... All things tend toward entropy. All things tend to disorder... all things fall apart.
>
> What's more, science has proved this in over a hundred years of research. Right?...
>
> Yes, the brand-new cosmos looks like randomness and entropy.... At first, all I see is a mixed up random flurry of protons and neutrons jittering maniacally in the scalding soup

of a plasma.... But... squint and take a look at the big picture... and you'll see something that makes randomness and
disorder look ridiculous... you'll see order on a level that
defies belief....

These gazillions of crashing particles are cooperating in
the formation of waves and troughs... that ripple from one
end of the cosmos to the other.... They are rippling as coher-
ently as ropes of clay, ropes that stretch across the cosmos for
hundreds of light-years, waves that roll protons and neutrons
in tight synchrony, waves that retain their identity until they
reach distant corners of the cosmos hundreds of thousands
of light-years from the point where they began.... And they
are so regularly and harmoniously—yes, harmoniously—
spaced that cosmologists call them musical.... [A]strophysi-
cists say this early cosmos and its plasma rang like a massive
gong....

Is this harmony of pressure waves, this symphonic spacing
of universe-spanning ripples, this mass choreography of ele-
mentary-particle pulses, entropy? Is it a tendency toward
disorder? Is it what... [the chemist] Frank L. Lambert calls
mere 'energy dispersal'? Is this entropy at work? No... it's so
anti-entropic that those in the scientific world who are trying
desperately to rescue entropy from the ubiquity of form and
structure call it 'negentropy.'

And then, as the eons pass in this daunting dance of the elements viewed
from his café table in Brooklyn at the beginning of the universe, Bloom
admits to boredom. He's been sitting there 379,000 years after all, and
things are starting to slow down. "At roughly the 380,000 year mark
after the big bang, the particles in the plasma [are] cooling." Everywhere
across the cosmos the quarks that have punctiliously clicked into six
protons for every neutron are now picking up relatively infinitesimal
electrons for the formation of the elements needed for stars.

Their fit is more precise than anything that even the makers of the ultimate high-precision scientific device, CERN's Large Hadron Collider, have ever been able to achieve.

If this were a truly random universe this fit simply should not be.... But our universe does not blat out more than a zillion to a zillionth power new forms of atoms, as the probabilistic equations of randomness would imply.... It produces just three rigidly constrained species of atoms. [Hydrogen, helium, and lithium all appear at the same time,] with astonishing supersynchrony.... It doesn't follow the rules of randomness....[2]

For their simplest examples, probability theorists tend to use dice, which have thirty-six possible outcomes when tossed around in a cup. "How can a universe of nearly infinite dice and nearly infinite tosses," asks Bloom, "produce just three varieties of atoms? This is staggering conformity and self-control... not mere trial and error...."

Then two billion years later, from seemingly unrelated supernovae in the explosive death of stars, emerged another gas, oxygen (Greek for "acid-maker"), together with all the carbon, nitrogen, and iron indispensable to life. That oxygen, when combined with hydrogen gas, not only produced a further explosion and a further gaseous substance, it engendered the properties of water that make possible our planetary home and bountiful bodies, in an astronomical cascade of singularities, from quarks to carbon to us.

"So what is it?" asks Bloom. "It's the paradox of the supersized surprise. It's the mind-snarler at the core of cosmic creativity. It is the question at the heart of the God Problem."[3]

I will leave the perplexing theological ruminations to Bloom. Physical entropy began with Clausius and Carnot, with steam engines and analogous energy-cycling mechanisms, where the law still dutifully applies. But these analogies are unhelpful in the social sciences, which are the domain of free human minds. They led Adam Smith to ponder fecklessly the

"Great Machine of the cosmos." It was the metaphor of the epoch. But the universe is not like a steam engine or any other kind of machine. It is not constantly subsiding into thermal equilibrium. It is an engine of ideas, an information system, like an economy.

The second law of thermodynamics fails because the universe is not statistical. It is a singularity full of detailed and improbable information. It is a "super-surprise." It is high in entropy from the outset. Information theory holds that it is in principle impossible to differentiate between a random sequence and a sequence of creative surprises. As Hubert Yockey, a founder of biological information theory, writes, "It is fundamentally undecidable whether a given sequence has been generated by a stochastic process or by a highly organized one."[4] All the evidence for a random universe is equally applicable to one full of information and creativity.

In economics, we are interested in complexity: the production of improbable arrangements of data and matter. This is the supply side. The measure of complexity is information content, gauged by the length of the computer program needed to produce it. The supply side, with all its intricacies of goods and services, commands far more information than the homogeneous money-denominated demand side. What Smith called the division of labor is the supply-side proliferation of information and entropy deriving from specialization and innovation.

The more random a sequence is and the longer the software program needed to generate it is, the more information it contains and the more complexity it manifests. On the scale of complexity, entropy, information, disorder, and apparent randomness are at the high end; low entropy, regularity, order, and low information content are at the low end. It runs from freedom of choice and surprise to determinism and predictability.

In information theory, *order* means lack of surprise and absence of creativity. This is understandable, since orderly sequences allow you to project future sequences from past ones. The pattern of order renders it intelligible and predictable. Disorder means unexpected and unpredictable outcomes in a sequence where previous bits of information do not determine subsequent bits. Examples include stock prices, profits, and interest rates, among other phenomena that have been demonstrated to not be "serially correlated."

The second law works by showing that the arrangement of particles in an entropic soup is less predictable and more disorderly than the arrangement of particles in a machine, such as a refrigerator or an internal combustion engine, which produces a usable potential difference between hot and cold, or usable voltages between electronic anodes and cathodes. But dwarfing this narrow insight into energy entropy is Shannon's entropy of information, surprise, complexity, and profit, which drives economic growth.

Most of economics has plighted its troth to concepts of order and disorder that profoundly conflict with what we learn from information theory. Perhaps the best economists come from the Austrian school, built on the work of Ludwig von Mises and Friedrich Hayek and led today by Mark Skousen and Roger Garrison. They have a vital grasp of the specificity and heterogeneity of capital and its interlocking relationships with entrepreneurship and skills. Capital is not a flow of investment or spending. It is not usefully gauged as demand. It is a complex and interlocking creative structure of production tools and skills evolved over time and full of information.

The Austrians have given us a model of macroeconomics called "spontaneous order." Nowhere does economics clash more deeply with information theory than on this popular theme. Cropping up everywhere from T-shirts and scientific texts to Ron Paul's presidential race and the deepest cerebrations at the Santa Fe Institute, "spontaneous order" is a direct descendant of Adam Smith's invisible hand, with a similar heritage in the natural sciences. There the concept appears under the guise of "emergence," a catchall term for events or patterns that cannot be calculated or precisely predicted.

The theory of spontaneous order maintains that complexity and equilibrium, such as are exhibited in economics, can emerge without planning or control, like biological ecosystems, planetary orbits, or human consciousness. Like "emergence," the concept enables scientists to "explain" orderly phenomena that could not have been predicted or causally specified from their background conditions.

Hayek developed the idea, conveying his belief that economics and culture evolve with the same unguided spontaneity as biological systems. Over subsequent decades, spontaneous order became a dominant theme of economic thought. The precocious Paul Krugman invoked it in *The Self-Organizing Economy*,[5] and it is found in Mormon philosophical blogs and Tea Party tracts. "I believe in spontaneous order" became a slogan of the right.

Even Keynesian economic models assume spontaneous order. Krugman and his followers believe that the responsibility of government is to push the economy toward a condition of full-employment equilibrium and growth. Most of the history of economics revolves around the issues of how order and equilibrium, partial or general, can be maintained.

Information theory finds all these concepts incoherent and self-contradictory. A capitalist economy epitomizes a complex dynamic system. In information theory, order is low-entropy. Characterized by regularity and redundancy, it tends to be predictable. It is low in information and surprise. Complexity, on the other hand, is high in information. It is high-entropy and the opposite of order.

When Adam Smith described a capitalist economy as "a great machine with every part adjusted with the nicest artifice to the ends they are intended to produce," he was depicting a low-entropy determinist order. Such an order could not generate novelty and innovation, high-entropy inventions and news. It could not produce long-term economic growth. Perhaps that is why Smith entertained the idea of capitalism evolving into a stationary state and why his first great followers were Ricardo and Malthus, both exponents of stasis as the destiny of capitalism. None of these economists of equilibrium and order was comfortable with a dynamic economics of permanent growth and change, disruption and surprisal.

Things that are growing and changing are by definition high in entropy. Moving from a settled past into an undetermined future, they are always defined by their information, their news, their surprises. Spontaneous order is self-contradictory. *Spontaneity* connotes the ebullition of surprises. It is highly entropic and disorderly. It is entrepreneurial and complex. *Order* connotes predictability and equilibrium. It is what is *not* spontaneous. It includes moral codes, constitutional restraints, personal disciplines, educational integrity, predictable laws, reliable courts, stable money, trustworthy finance, strong families, dependable defense, and police powers. Order requires political guidance, sovereignty, and leadership. It normally entails religious beliefs. The entire saga of the history of the West conveys the courage and sacrifice necessary to enforce and defend these values against their enemies.

A key principle of information theory is that it takes a low-entropy carrier to bear high-entropy creations. All the surprising singularities of creative capitalism depend on the boring regularities of political order. Maintenance of the low-entropy carrier cannot be left to some imaginary spontaneous order.

Even though the universe displays a providential pattern, the spontaneous trend of low-entropy carriers on earth is to deteriorate, as the second law would ordain. Politicians and regulators make high-entropy interventions for the interests of their cronies. Lawyers exploit the law for the advantage of the bar. Incumbent legislators punish those who fund their challengers. Tax rates creep up, stultifying learning curves and destroying the information fabric of entrepreneurship. Mobs rise up in envy and indignation to attack successful ethnic or religious groups, like Jews or Mormons, as well as "the rich."

The key misconception of the popular versions of libertarian and Austrian economics is that political order can be spontaneous—that capitalism can thrive in anarchy. But central to the Austrian model is the power of prices for signaling economic conditions. Without the government's enforcement of property rights and contracts and its maintenance of defense and a monetary system, the carrier fills up with noise.

Information theory defines entropy as freedom of choice and surprise. It is intrinsically libertarian in its implications. But it does not presuppose libertarian anarchy. Legal codes and moral practices do not spring spontaneously from the interplay of the marketplace or from the self-interested conflicts of individuals. Progress in law and order does not spring from a Darwinian process of natural selection among random mutations. It is achieved through a heroic struggle to develop civilized institutions and defend them. Building these institutions are leaders, political and intellectual. The quality of political leadership is crucial to the development and defense of markets, and there is nothing spontaneous about it.

The low-entropy side of the economy is demand and predictability; the high-entropy side is supply and surprise. Government and law are on the low-entropy side; they favor and foster rules of order. On the high-entropy side are entrepreneurship and spontaneity, the domains of creativity and surprise. Spontaneous order is an oxymoron that violates the fundamentals of any information system.

This error of economics precedes information theory. Hayek struggled with its implications throughout his career. On some occasions, he spoke of spontaneous and self-organizing cultures and legal systems evolving from the bottom up without hierarchical guidance. At other times he was an earnest exponent of the need for deliberate constitutions and moral mandates. Addressing the issue, he wrote the *Constitution of Liberty*.[6] But America's libertarian tradition has embraced chiefly the spontaneity of order; its origins in an evolutionary bottom-up process; and its resistance to top-down cultural controls or influences such as religion, family structure, and legal constraints.

Entropy is Janus-faced. Its upside surprises are redemptive and favorable to freedom. It is freedom of choice. But the carrier itself requires constant vigilance against entropic noise.

Order is not spontaneous, but it is a necessary condition for all the surprises of freedom and opportunity.

Romer's Recipes
and Their Limits

A FEW YEARS BEFORE William Nordhaus published his historic study of technology and the cost of lighting, a thirty-five-year-old economist published a paper that brought information and knowledge to center stage in theoretical economics. Paul Romer—whose father, Roy, was a governor of Colorado and became a co-chairman of the Democratic National Committee—had an undergraduate degree from the University of Chicago in physics, not in economics, the field in which he would make his mark. "Not an avid member of any clan," as David Warsh describes Romer, "he walked out of the two best departments in the discipline, Chicago's and MIT's."[1]

Romer—now one of the leading economists in the world—looked at things differently from his brethren in Chicago and Cambridge. "People acted as if economic analysis couldn't help us understand why the rate of technological change might be speeding up," Romer told the economist Arnold Kling of the Cato Institute. Romer himself was thinking, "This may be the most important question in human history."[2]

In the fall of 1989, with technological change on his mind, he took a leap of faith—he left the tenure track at Chicago to be near Silicon Valley. He picked up a one-year fellowship at an institution funded by the technological change of a century ago: the Ford Foundation's interdisciplinary Center for Advanced Study of Behavioral Science ("behavioral" being Henry Ford's euphemism for "social," a word that sounded too much like "socialism").

Romer's leap landed him on terra firma. By late 1990, he had a job at Berkeley and had published his now-famous paper, which would establish him as a pioneer of a movement called New Growth Theory. In contrast to his predecessors, who treated technological change as "exogenous" if they treated it at all, Romer titled his paper "Endogenous Technological Change." Since technological change is the driving force of economic growth, moving innovation from outside the scope of economics to the core of the discipline is crucial to the redemption of the field. It was the culmination of Romer's ceremonious but firm unseating of the four horsemen of economic muddle—self-extending markets; static input-output tables (without innovation); reified demand; and, at the heart of the fogbank, equilibrium.

As Kevin Kelly commented in a profile of Romer six years later in *Wired*, "economists, like cultists awaiting the apocalypse, had long looked to the day growth would end."[3] They realized, on some level, that "the set of traded goods in an economy is always changing," but, as Romer commented acerbically in 1993, in their models, "this turbulence is an epiphenomenon of no fundamental interest."[4]

Romer looked where the cultists were looking and saw no signs of apocalypse: "I've been an optimist ever since I got started in economics," he told his fellow economics professor Russ Roberts in 2007. "It may be just a personality trait but I think it's been reinforced by the research." He had been a graduate student in the late 1970s, "back in a time when people talked about the limits to growth....People were saying that our standard of living ... was going to collapse—there was no way we could sustain it. Those kinds of pessimistic forecasts have been made ever since the time of Malthus. And they've always been wrong." Echoing the apostles of the learning curve, from Henry Adams and Bruce Henderson to William Bain and Ray Kurzweil, Romer declared, "The historical pattern has been one of accelerating growth—not just sustained growth but accelerating growth."[5]

Romer's research led him through mathematical thickets and beyond the economist's traditional trinity of land, labor, and capital to a model of growth that depended on technological change. Romer stressed not merely the calibration of labor costs and capital spending in a "production function," but the transformative effect of ideas.

"I think part of why this question attracted me was because of my background in physics, and to a physicist, the whole notion of a production function sounds wrong. We don't really produce anything. Everything was already here, so all we can ever do is rearrange things. Think of conservation of mass. We've got the same amount of stuff we've always had, but the world is a nicer place to live in because we've rearranged it…."[6]

From this classical physics point of view, Romer thought about the structural or chemical changes that make up that "rearranging." He realized, "It's like cooking." And here, Romer proposed, is where creation comes. Milk and other ingredients can be artfully brought together to "create something—a soufflé, which is really valuable, and gives us great pleasure when we eat it." Clumsy or ignorant rearranging, on the other hand, leads to something worth less than what you started with: "sour milk."[7]

Beyond the skill of the chef and the quality of the ingredients and equipment, a good recipe itself is a valuable thing: "A canonical example is turning sand on the beach into semiconductors." This, Romer believed, is how people create value in an economy—not burning through scarce raw materials until they're all gone, but creating endless recipes to rearrange what's there into states of greater and greater value. As a result, "there is absolutely no reason why we cannot have persistent growth as far into the future as you can imagine."[8]

So, by 1990, at least a few economists were grudgingly admitting that technological change stems from human creativity—from ideas. Romer took another step closer to information theory as he continued to mull over the question of the acceleration of technological growth.

"Evolution has not made us any smarter in the last 100,000 years," he said to Ronald Bailey (in an interview in *Reason* titled "Post-Scarcity Prophet").

> Why for almost all of that time is there nothing going on, and then in the last 200 years things suddenly just go nuts?
>
> One answer is that the more people you're around, the better off you're going to be.... If everything were just objects, like trees, then more people means there's less wood per person. But if somebody discovers an idea, everybody gets to use it, so the more people you have who are potentially looking for ideas, the better off we're all going to be. And each time we made a little improvement in technology, we could support a slightly larger population, and that led to more people who could go out and discover some new technology.[9]

But some countries had lots of people and not very much technology, so there had to be another piece to the puzzle.

> Another answer is that we developed better institutions. Neither the institutions of the market nor the institutions of science existed even as late as the Middle Ages. Instead we had the feudal system, where peasants couldn't decide where to work and the lord couldn't sell his land. On the science side, we had alchemy. What did you do if you discovered anything? You kept it secret. The last thing you'd do was tell anybody.[10]

But institutions are like ideas—they can be discovered, and once they are discovered, they can be shared: "When good institutions work somewhere in the world, other places can copy them.... New Growth Theory describes what's possible for us but says very explicitly that if you don't have the right institutions in place, it won't happen."[11]

In other words, the high-entropy signal will never reach its destination if it doesn't have a low-entropy carrier.

Crucial to Romer's scheme of knowledge-based growth is the idea of rival and non-rival goods with varying degrees of excludability. The distinction was evident to Thomas Jefferson in the early nineteenth century, to Henry Adams in the late nineteenth century,[12] and was upgraded for modern use by a number of writers before Romer—including Nicholas Negroponte, who distinguished between sharable bits and consumable atoms.[13] Perhaps Jefferson was most eloquent: "He who receives an idea from me receives instruction himself without lessening mine; as he who lights his taper at mine receives light without darkening me.... Ideas are incapable of confinement or exclusive appropriation."[14]

Romer can be credited with more cumbersome language, more mathematical symbols, and a willingness to put up with the considerable demands of the economics fraternity. But his invaluable contribution was to integrate the distinction with all his other insights in a model that ingeniously advances the profession into the information age.

A rival good is a thing that can be appropriated by only one person at a time—an egg, an apple, a book, a bond, a tennis racket, or an apartment. It tends to be used up as it is used. Non-rival goods, by contrast, are appropriable by any number of people at one time. As you use non-rival goods and services, they expand, according to network effects (Metcalfe's Law), by the square of the number of compatibly linked users. Examples are books on Kindle, Google searches, Quicken spreadsheets, operating systems, dress designs, songs, television programs, or economic ideas.

Romer still tends to treat ideas like matter and to confuse order with information; still, his math has led him close to information theory. As his career developed, he continued to focus on how knowledge reaches people. In 1993, he published a study on the economy of the tiny Indian Ocean island of Mauritius, which for a century had maintained a wall of tariffs to foster local industry—to little effect. In 1970, the government created a special enterprise zone, and soon a third of all the jobs on the island were in garment-making firms headquartered in Hong Kong.

Knowledge is necessary to break into an enterprise, and it is often latent in the companies themselves. "This knowledge did not leak in from Hong Kong," wrote Romer. "It was brought in when entrepreneurs were presented with an opportunity that let them earn a profit on the knowledge they possessed."[15]

What happened in Mauritius was happening elsewhere; it was "one of the untold stories about the '80s and '90s," Romer told Ronald Bailey. "If you track the legislative history on foreign investment, you see a colonial legacy, even as late as the '70s, where developing countries have laws designed to keep corporations out. Then there's this dramatic turnaround as they saw the benefits that a few key economies received by inviting in foreign investment.

"It's not the people from the developing world who are making the argument that Nike is a threat to their sovereignty or well-being. It's people in the United States," Romer said. He believed in rational action: "The people in the developing world understand pretty clearly where their self-interest lies."[16]

The fact that a worker in a Nike factory in Vietnam has a lower quality of life than a worker in the United States "feels wrong to many of us," Romer said in another interview, "but that's not the question here. The question is, did Nike's coming in make the life of that person better off or worse off? The unambiguous answer is that Nike coming in really helps that person and helps many other people in that country."[17]

When faced with irrationality, Romer sometimes ignored it and sometimes got frustrated: "Just look at the facts. The protestors are amazingly ignorant about what has happened in terms of, say, life expectancy. Life expectancy for people in the poorest countries of the world is now better than life expectancy in England when Malthus was so worried about it" spiraling into unsustainable masses of population.[18]

Romer often spoke of the equal importance of the institutions of science and the market, but he was feeling more and more frustrated with academia. In 1996, he started teaching at Stanford's business school, one step closer to the market itself. "I wanted to not just talk the talk from an economics department where I could ignore students and tuition. I wanted to come live and work in a business school and see if that really is the way more of higher education should look."[19] The next year, he was named one of America's most influential people by *Time* magazine. ("A sage for the silicon age, Romer is upgrading the dismal science to keep pace with the digital revolution."[20])

Despite his move to business school, he still felt that the institution of science was lagging behind the institution of the market in effectiveness. While many students passionately studied subjects unlikely to give them jobs—and thus allow them to help others—upon graduation, many others in the "practical" subjects, including the sciences, weren't spending enough time struggling with the material to really learn it. Romer's solution was technological: an online program, tailored to individual textbooks, which questioned, guided, and tested students, allowing their professors to focus on teaching rather than on grading. In 2000, this idea culminated in an online company, Aplia, and by 2009, the *Washington Post* reported that a quarter of all economics students were working with Romer's program.[21] His father, meanwhile, was taking on education at a different level as superintendent of schools in Los Angeles County, a daunting job.

As Roy Romer fought for greater access for needy students to more functional schools across Los Angeles, Paul Romer was thinking about a related problem on a global level: how needy people looking for work could have access to more functional cities around the world. For that, the sticky institutions of science and education could wait; the crucial thing was the low-entropy carrier of the market. "For the purposes of thinking about development, it's probably not a bad shortcut to say that the only institutions that matter are the institutions of the market," Romer told Arnold Kling. "Get property rights, the basic institutions of security, personal security, a legal system that supports property rights—get those in place and things will be fine."[22]

But for poor countries with long histories of upheaval and erratic government, that low-entropy carrier—which has to be persuasively permanent—seems impossible. Romer's thoughts were coalescing around an idea of foreign-run "charter cities" in poor countries, an idea that grew out of the history of his own career. Technological change improves the quality of life; the institutions of science and the market allow technological change to occur; stability, free minds, and free trade allow science and the market to develop; and the more people involved, the better for all. These charter cities would import their low-entropy institutions—laws and market—from a stable foreign power, giving other stable foreign powers the confidence to invest. Jobs could be created and people could gain skills, bettering their own lives and the lives of people near and far as human minds were freed from grinding poverty through interconnectivity with the world.

Romer likes to point out a haunting photo on the main page of the Charter Cities' website: a row of well-dressed schoolboys sitting on concrete stumps under the streetlamps of the airport parking lot in Conakry, Guinea, doing their homework. He notes that they likely all have cell phones, but the price controls Guinea imposes on utilities mean they do not have electricity at home. The issue is not money or motivation; it is, as Romer says, "rules." If Guinea had a system that did not stymie investment in infrastructure, the technology would come. Romer's charter cities idea was a way of cutting the Gordian knot of entrenched, dysfunctional regimes that preclude genuine development. His dream is that the successful charter city could infect its host country with its low-entropy carriers. And he can point to a powerful precedent. As Romer told the *Atlantic*, "Britain inadvertently, through its actions in Hong Kong, did more to reduce world poverty than all the aid programs that we've undertaken in the last century."[23] Hong Kong infected China with "Free Zones" that became manufacturing centers for the world.

The problem with Romer's charter cities idea is that it sounds too much like imperialism for some people. The emotional response of anti-imperialism has kept charter cities from getting off the ground anywhere they've been tried. Two governments that have considered charter cities have run into trouble: the president of Madagascar was brought down in a coup in 2009 over charter cities, and the president of Honduras is facing stiff resistance to the idea of foreign encroachment. As Romer likes to tell his students, "Everyone wants growth but nobody wants change."

Still, change keeps coming, fueled by technology, as Romer's 1990 paper reminded the economics fraternity. As productivity grows, technology keeps freeing people. And this, as Romer notes, "really is, in some sense, the scarcest commodity: the power of the human intellect."[24]

Romer comes up short only in his definition of entrepreneurial creativity. Romer's summary of what entrepreneurs do is "they reassemble chemical elements." Indeed they do, but so does a lightning strike or a hurricane, sunlight or a tsunami. None of those has anything to do with entrepreneurial creativity.

Entrepreneurial creation is the generation, de novo, of novelty and surprise—freedom of choice originating in the world of ideas, and imagination beyond all concern with chemicals. The contrary view—that all ideas are determined by material relationships—is the materialist superstition.

Academic scientists of any sort expect to be struck by lightning if they celebrate real creation de novo in the world. One does not expect modern scientists to address creation by God. They have a right to their professional figments such as infinite multiple parallel universes. But it is a strange testimony to our academic life that they also feel it is necessary to deny creation to human beings. With his attempted reduction of entrepreneurship to chemistry and cuisine, Romer finally succumbs to the materialist superstition: the idea that human beings and their ideas are ultimately material.

Out of the scientistic fog there emerged in the middle of the last century the countervailing ideas of information theory and computer science. The progenitor of information theory, and perhaps the pivotal figure in the recent history of human thought, was Kurt Gödel, the eccentric Austrian genius and intimate of Einstein who drove determinism from its strongest and most indispensable redoubt: the coherence, consistency, and self-sufficiency of mathematics.

Gödel demonstrated that every logical scheme, including mathematics, is dependent on axioms that it cannot prove and that cannot be reduced to the scheme itself. In an elegant mathematical proof, introduced to the world by the great mathematician and computer scientist John von

Neumann in September 1930, Gödel demonstrated that mathematics is intrinsically incomplete. Gödel was reportedly concerned that he might have inadvertently proved the existence of God, a *faux pas* in his Viennese and Princeton circles. It was one of the famously paranoid Gödel's more reasonable fears. As the economist Steven Landsberg, an academic atheist, put it, "Mathematics is the only faith-based science that can prove it."[25]

For the purposes of economics, however, who or what occupies the top of the hierarchy of existence is less important than the human judgment and creativity below. The demonstration that any logical system must rest on un-provable axioms opened the way for the autonomous creations of human beings. The logical system could never subsume the free mind that launched it.

The Promethean computer theorist Alan Turing continued the dismantling of deterministic scientism. In conceiving a computer architecture that is universal—the Turing machine—he extended the Gödel proof to computer science. Computers are embodiments of mathematical reasoning, and Turing had invented the quintessential determinist machine. Now he went on to show that no computing machine could prove all the principles on which it was based—that a Turing machine could not even calculate when or whether any particular program would halt and deliver an output.

Turing had revealed that any program could be encoded in numbers. He and his followers—Emil Post, Alonzo Church, and Gregory Chaitin, among others—demonstrated that most numbers are incomputable: they cannot be calculated by a computer program. Whatever the universe may be, it is not a mechanistic logical machine. It requires axioms beyond itself.

In his doctoral thesis, published in 1939, Turing elaborated on the implications of Gödel's insight: "Gödel shows that every system of logic is in a certain sense incomplete, but at the same time ...indicates means whereby from a system of logic ...a more complete system ...may be obtained" by incorporating its necessary exogenous or outside axioms. Then a further system could be contrived that included the first one together with its enabling presuppositions, and so on. Turing imagined a deterministic computing machine that made non-deterministic leaps when necessary by consulting "a kind of oracle as it were. We shall not go any further into the nature of the oracle apart from saying that it cannot be a machine."[26]

George Dyson tells the story in his definitive history, *Turing's Cathedral*. "The Universal Turing Machine of 1936 gets all the attention," Dyson writes, "but Turing's O-machines [oracle-machines] of 1939 may be closer to the way intelligence (real and artificial) works." As Turing explains, "Mathematical reasoning may be regarded rather schematically as the exercise of a combination of two faculties which we may call intuition and ingenuity. Intuition consists in making spontaneous judgments, which are not the result of conscious trains of reasoning. These judgments are often but by no means invariably 'correct.'"[27]

Dyson goes on to tell the tale of Turing and the often-intuitive cryptography that broke Germany's Enigma codes at Bletchley Park and may well have been decisive in winning World War II in Europe. But Turing's O-machine could also be depicted as an economic model. The economy is like an O-machine in which entrepreneurs play the central oracular role. In launching logical or mechanistic systems, they are not merely uncovering new domains of a hermetic and universal logical and material order. They are making autonomous inventions or creations that enable new logical systems: machines.

In *Wealth and Poverty*, I criticized the principle of "look before you leap":

> [It] misses the necessary bound of surprise in all radical innovation. The idea that businesses buy knowledge like any other factor of production until its cost exceeds its yield, that businesses can safely and systematically assemble facts until the ground ahead stretches firmly before them, misses the radical difference between knowledge and everything else. It is the leap, not the look, that generates the crucial information; the leap through time and space, beyond the swarm of observable fact that opens up the vista of discovery. Galileo broached the modern age of science not by observing thousands of factual

trajectories and deriving from them the law of gravity; rather "I conceived as the work of my own mind a moving object launched above a horizontal plane and freed of all impediment." Freed, that is, of the facts; freed, by a leap of imagination, of the conditions of all real moving bodies as they are buffeted through the resisting air. Imagination precedes knowledge. Creative thought is not an inductive process in which a scientist accumulates evidence in a neutral and "objective" way until a theory becomes visible in it. Rather the theory comes first and determines what evidence can be seen.

In the capstone of the new information theory, Shannon took the key further step of showing that information is not order but surprise. Deterministic sciences had always relied, as they do today, on the proposition that information is order and order is ultimately expressed by hermetic mathematical logic manifested in the material environment.

In every case, the material environment is the determining reality. The goal of science, in its scientistic guise, is to grasp so fully the nature of material conditions that all important events, such as the human mind or the universe, are explained by the encounter of random fluctuations of matter with physical laws: the Brownian motion of particles filtered by the laws of physics and chemistry. This theory entails an act of intellectual legerdemain like the disappearing infinitesimals of the calculus; human knowledge ends by mastering the natural order and then succumbing to its evanescence in entropy according to the Second Law of Thermodynamics. Physical entropy dictates that all voltages and other potential differences that yield power will dissipate to equiprobable states of random motion. Knowledge gives way to the chaos from which it sprung. In this vision, information is order, equilibrium, determinism, and predictability.

Shannon overthrew this entire scheme. Information, he established, is entropy. It is disorder, deformations of order, disruptions of equilibrium. It is indeterminism and surprise. And entropy is freedom of choice. This insight is at the heart of the information theory of capitalism. Serious economists can no longer ignore it.

Mind over Matter

EVERYONE THINKS HE knows what information is. Information is what we know. Information is change in what we know. Information is how we communicate. Information is on the Internet. It is growing exponentially. It is data or messages or media or music or money or recipes or websites or pixels or packets. It is memory. It is imagination. Information is in the cloud. It is in the air. It is weather or *Weltanschauung* or winks across a room. It is shrugging shoulders or pointing fingers. It is bits and words and bytes and gigabytes and zettabytes. It flips and flops, on and off, everything and nothing.

Through the middle of the twentieth century, however, scientists could describe information no more definitively than they could define matter. In 1948, Shannon changed all that. He began by defining communication as the problem of reproducing at one place and time a message composed at another place and time and sent through the "noisy channel" of the material world: the conduit. It could be wire or air or time or space. The medium did not matter. Call it the universe.

The original measure of uncertainty of the information flow was the size of the alphabet or vocabulary available to the composer of the message. Information flow was highest when the alphabet or vocabulary was largest and the bits were most surprising, most unexpected, and most improbable. More information flowed, paradoxically, when the bits were random—that is, completely unpredictable. Entropy is freedom.

Upending the assumption that order and determinism are the foundation of any science and identifying the role of randomness and noise, Shannon made use of an insight of quantum theory. The determined universe gives way to a stochastic universe (from the Greek *stoche*, "chance"), shaped but not determined by low-entropy (predictable) laws that do not change.

Social scientists and biologists have attempted to stand aloof from the new imperium of information. As late as 2001, James Watson claimed that DNA demonstrates that genetic biology is "nothing but chemistry and physics."[1] But as Shannon understood, biology too has become another information science. As Bill Gates puts it, the genome is a software program but "far more complex than any we could build." The sugar phosphate backbone of DNA is the low-entropy carrier for the high-entropy messages of the genome for all creatures over all time. The "central dogma" of Watson's own collaborator, Francis Crick, holds that the information in the genome can define the composition of proteins but proteins cannot inform or change the content of the message in the DNA bases. The word comes before the flesh.[2]

Thus information theory refutes the inheritance of acquired characteristics. The Soviet biologist Trofim Lysenko thought that physical changes in proteins in life could be genetically blended and transmitted to future generations. Stalin anointed Lysenko's views as the official Soviet biology because they supported the notion of a "New Soviet Man" freed of the influence of his ancestors.[3]

The new biology and computer science show that the information present in the genome is a software program that is independent of any particular carrier or substrate. The inventor-futurist Ray Kurzweil, therefore, can conceive of uploading his carbon-based genome into the silicon memory system of a supercomputer. Compugen and Evogene can take the DNA cascade out of its organic carbon realm, with all its intricate cellular chemistry, redundancies, and feedback loops, and make in silico models to perform predictive bioscience in silicon-based computers.

Information and matter emerge as complementary but intrinsically distinct. As MIT's Norbert Weiner wrote in 1948, the year that his rival Shannon launched information theory, "Information is information, not matter or energy. No materialism that does not admit this can survive at the present day."[4] The laws of matter may be low-entropy and largely determined. But information is surprise. Enterprise fuses the low-entropy carriers with the high-entropy creations.

Information theory, in its foundational role in computer and communications science, has been crucial to the revitalization of the American and world economies. The works of Kurt Gödel, Alan Turing, John von Neumann, and above all of Claude Shannon created the architecture of ideas and systems underlying and informing digital computers, fiber optic networks, wireless communications, and the global Internet. Shannon's insights explain why all the high-entropy information in society is migrating to forms of spectral light: the electromagnetic spectrum of frequencies guaranteed by the speed of light. This ultimate low-entropy carrier is manifested in wireless signals through the air and in photonic signals through fiber optic threads around the globe. The same ideas and systems are behind the emergence of "cloud computing" and big data mining. The predictable low entropy of Shannon's formulae enables all the sorting, searching, and pattern matching amid the blizzards of new information from the entrepreneurial laboratories of information industry.

Shannon's vision, however, is not merely a set of ideas behind the "computer revolution." It is also supplying a new epistemic foundation— a new way of thinking about knowledge and power—for all the sciences, from physics and biology to psychology and economics. Central to all science is the free human being who conceives and defines it. As Einstein explained, "The belief in an external world independent of the perceiving subject is the basis of all natural science."[5] Without conscious, creative, free, and powerful human minds and wills, science is meaningless and impossible. The most crippling force in science is therefore the pervasive effort to define man as the determined product of material forces. This campaign, should it succeed, would destroy the culture that sustains science. At its heart is an anti-scientific philosophy—not real science but its perverted cousin, *scientism*.

This philosophy afflicts every branch of science. Biologists reduce man to a probabilistic function in a mechanistic scheme of natural selection. Neuroscientists reduce mind to material fluctuations in the brain. Evolutionary psychologists reduce learning to habit-formation through stimulus and response. Physicists are bewitched by the bizarre and stultifying anthropic principle of infinite multiple parallel universes, a materialistic blow to the indeterminism of quantum theory. They all have made their sciences into systems of order devoted to subduing surprise.

• • • •

They were a wriggling, slithering, writhing, squiggling,
raveling, wrestling swarm of maggots rooting over and
under one another in a heedless, literally headless, frenzy
to get at the dead meat. She learned later that they were
decephalized larvae. They had no heads. The frenzy was
all they had. They didn't have five senses; they had one,
the urge, and the urge was all they felt....
Just take a look at them....the billionaires![6]

—TOM WOLFE, BACK TO BLOOD

That's how Tom Wolfe describes America's one-hundredth of 1 percent—billionaires reduced to primal itches and urges as they grapple their way toward the exhibits at an auction in Miami of outlandish contemporary art.

Wolfe's genius is to capture in one suppurating and sordid image the precise vision of humanity that ultimately animates nearly all the prevailing sciences, from biology to psychology, and from sociology to economics. At the center of all these theories is an image of man. It goes by many names, from id to animus, function to factor, random walker to hedonic maximizer, *homo economicus* to *homo sapiens*, consumer to entrepreneur. But in the end, aggregated and subsumed in the mathematics of the model, it is always the same randomly wriggling worm, mostly devoid of free will, intelligence, imagination, or even consciousness. All the model has to drive it is the *urge*.

The philosopher Jerry Fodor and the Italian biologist-cum-cognitive-scientist Massimo Piatelli-Palmarini[7] have arrived at a profound critique of the reductionism at the foundation of all the social and life sciences. In language considerably less colorful than Wolfe's, they show that most academic analyses of living beings rely on a heuristic simplification that leaves out creativity and mind. They coined the ungainly acronym IEGMLM to identify six recurrent assumptions in these disciplines.

These six ideas are all jawbreakers or mindbenders, but they are worth considering, as they are the bricks in the foundation of the materialist superstition. The first is *iterativity*. Whether we're talking about evolution of learning from infancy or evolution of species from a primal soup, it's a random walk all the way, starting and continuing under the same rules from any point and always remaining open to further evolution or conditioning.

The next is *environmentalism*, but not what you think. In IEGMLM, environmentalism means that nothing matters in the iteration but actual interactions with the immediate physical surroundings.

Third is *gradualism*: advancing only by "numerous successive slight modifications," as Darwin put it. Monkeys don't jump over the moon, or land on it—even in their minds, if they are deemed to have them.

Fourth is the important concept of *monotonicity*. This means one unchanging scheme of cause and effect. The iterations are always and everywhere chiefly governed by the same single factor, which always follows the same rules, whether reward in learning or selection of the fittest in biology. With an adequately detailed theory of learning, as one eminent Skinnerian opined, "we could teach English to worms." With a sufficiently extensive model of recursive natural selection, repeating itself on itself, we could evolve a lump of clay into Arnold Schwarzenegger. As Fodor and Piatelli-Palmarini comment, "These are very strong claims. In real life ... practically nothing is a monotonic function of practically anything else," even Arnold. The world is full of agents who change their responses or reverse them.

A related concept is *locality*, which means insensitivity to remote influences, past events, future events, or probabilities. To the agent in Dubuque, China doesn't matter. Nor does the Holocaust. Nor does Omaha. Nor does next week. Locality leads to the last characteristic: *mindlessness*—no mental causes, intelligence, imagination, design, will, or knowledge allowed.

There you have it. In other words, IEGMLM is a mouthful of philosophical cogs, pins, widgets and chemical elements. In information theory, these factors represent an "ergodic" universe, which assures that the same functions will reliably produce the same outputs. Ergodicity is crucial to any predictive model. If the model itself generates a variety of outcomes, determinist prediction is impossible.

The IEGMLM principles govern the functions of a universal black box, inside of which is nothing but a random generator followed by a test or filter system. In biology and behavioral psychology, there is a producer of traits and a filter of the traits. In this scheme, learning is merely the forming of habits. Evolution is merely the filtered accumulation of the accidents of fitness.

In identifying the intrinsic logical unity between behavioral psychology and Darwinian biology, Fodor and Piatelli-Palmarini follow the path of the polymath philosopher-mathematician David Berlinski, who preceded them with several books. In effect, behavioral psychology is the micro-theory of evolutionary biology. It shows the contents of the learning steps taken by the members of species as they respond to their environment. In the contests of survival, the fittest make it through the filter. The rest do not. In neither case do intelligent entities act purposefully and willfully in the world; in general the world acts in them and through them.

So at the bottom of all these models is still the *urge*. It is the power that drives the system ahead, pushes the samples through the fitness filter, animates the random generators of behavior or the random initiatives of bioforms. In the end powering the urge is *energy*, whether in the form of electromagnetism or nuclear emissions or chemical ions or thermodynamic gradients of physical entropy. The destination of all the learning and evolution, according to the prevailing science, is the heat death of the universe, as all potential differences and gradients dissolve. The canonical story is that matter is all there is and in the long run we are all dead.

Because these principles deny such vivid and vital human experiences as consciousness, memory, imagination, and free will, no one actually believes in the model in all its implications. In the face of constant empirical disproofs, psychologists have abandoned Skinner's behaviorism, which has proved entirely unable to explain human acquisition of language, vision, motor skills, and other cognitive or even reflex functions. As a result, universities are teeming with such anti-Skinnerian disciplines as cognitive psychology, neuroscience, and bioinformatics. Examine them closely, though, and you see that these new fields simply shift the materialist model from external human behavior to the interactions of neurons and synapses across the dendritic spreads of the physical brain.

Meanwhile the essential incredibility of the materialist superstition—it is intractably counter-intuitive and even self-refuting, with mind denying mind—only enhances its grip on the scientistic imagination. Even though no one seriously thinks that man can be explained as a risen worm, the vision survives for its instructional uses. It is thought to capture some inner essence of human animality or material determinism.

The materialist superstition is most crippling, however, in economics, where it shapes government policy and real life. It is summed up in *homo economicus*, "economic man," the marginally upgraded and quantified IEGMLM function residing at the heart of nearly all economic models. *Homo economicus* is iterative (a random walker). His algorithm is a function of stimulus from his material local environment, and his learning process is governed by physical pleasure or pain, reward or punishment, profit or loss, in a monotonic hedonic path to power.

Relentlessly seeking equilibrium and order, *homo economicus* does not jump or duck; he does not create new things or leap ahead purposefully. He looks immediately ahead and moves toward the warming light, responding through his senses to his local surroundings, smoothing out all gaps and gradients, disequilibria and disruptions of prices and markets. Like the Skinnerian learner or the Darwinian species, he is a filter of exogenous or outside forces. He is restricted to Steven Johnson's "adjacent possible."

In economics, the passive consumer or optimizing worker may become an entrepreneur. Defined as a scout of physical opportunities, an arbitrageur of diverging prices, or Paul Romer's "assembler or re-assembler of chemical elements," the entrepreneur still is safely ensconced in his material environment. He still consists chiefly of chemical elements himself and seeks to restore equilibrium in his surroundings. By banishing real creativity, the model eliminates all new and surprising information.

Skinner's behaviorism failed in psychology, however, and it fails in biology. Thus it is even more misleading in economics, which translates psychology onto a collective plane.

Economic activity is not iterative or ergodic; it constantly changes. It is entropic, full of surprises. Businesses can never assume that doing the same thing—producing the same goods and services for decades—will ensure survival. Economic activity is not dominated by the material environment; its purpose is often to transform its environment through the application of new ideas and breakthroughs. It is not gradualist; it makes huge speculative and inventive leaps. It shows little monotonicity—what works on one occasion might fail on another. It is not local in time or space; it is not restricted to the so-called "adjacent possible"; its enterprises reach around the world and are deeply affected by remote developments and projected outcomes.

The power in capitalism must not be mindless. Unless it is combined with knowledge, mere economic power or money is fruitless. Enterprise involves memory of the past and anticipation of the future, and it is creative. It is not a simple incentive system of rewards and punishments, of carrots and sticks. It is an information system, and it is governed less by economic theory as we know it than by information theory. The beautiful congruence of information theory with a capitalist economy, the creative convergence of knowledge with power, has been the main subject of this book.

Knowledge is about the past. Entrepreneurship is about the future. We are connected to the past by our memories and to the future by our choices. Information theory moves from the future to the past, while physical theory moves from the past to the future. Events are determined by physical causes from the past and by subjective choices from the future. The entrepreneur surfs the crests of creation in between.

However convenient for models of the economy, the weather, and perhaps even the human body, determinism is inimical to a universe governed by information, which is defined as surprise. The determinist view in economics seeks perfect competition, perfect knowledge, and perfect predictability. A determined economy, Adam Smith's "great machine"—whether governed by an invisible hand, a spontaneous order, a Marxian dialectic, or an evolutionary mandate—has no surprises. The masters of such an economy are content with the simulacrum of enterprise in supposedly "free markets," but in reality there is no room for creative entrepreneurs or agents of change. Economic actors respond robotically to the random flux of their surroundings. It is an IEGMLM economy without capitalist creation.

Only the low entropy carrier of reliable law can enable the upside surprises of creative capitalism. Profits are the measure of the high entropy of innovation and invention. Profits thus are the fruits of disequilibrium and disorder, of entropy and freedom.

When law becomes high-entropy, the flow of information halts and knowledge withers. The line of demarcation between conduit and content dissolves, and all becomes noise. When currencies float chaotically and financial rules gyrate unpredictably, finance becomes a high-entropy and high-profit endeavor. The economy lurches toward a hypertrophy of finance and litigation.

At a time when conservatives and liberals, libertarians and authoritarians battle in the arena of ideas, information theory offers a redemptive synthesis. The low-entropy carriers that conservatives uphold of law, property, family, and morality enable the high-entropy creations of science and entrepreneurship. Information theory tells us that order is not spontaneous; information is not perfect; playing fields are not level; property rights and human rights are not automatic. They must be upheld and fulfilled by culture, religion, and politics. But information theory is also a mandate of liberty, enshrining freedom of choice as its deepest law of entropy and creativity.

PART TWO

The Crisis

The Scandal
of Money

AUSTAN GOOLSBEE, the chief economist in the Obama White House at the time of the 2008 financial crisis, commented, "It's almost like these guys should have gotten the Nobel Prize for evil." He was not speaking of Al Qaeda or Hamas but of the senior executives at AIG, the global insurance and reinsurance firm headed for thirty-eight years by the entrepreneurial titan Maurice "Hank" Greenberg.

In the fateful middle years of the first decade of the millennium, during the run-up to the so-called Great Recession, the AIG financial products group had issued a huge number of quasi-insurance policies called credit default swaps—some $2.7 trillion worth of coverage if all the insured securities failed. Coincidentally, the CEO, Hank Greenberg, was ousted from management during this period amid a fusillade of charges from New York Attorney General Eliot Spitzer, then on a hurried path to becoming governor of New York. Although his prosecutorial probe of AIG missed the $2.7 trillion, Spitzer did descry, amid the company's some $100 billion of global operations in 130 countries, a number of arguably deceptive accounting practices. These added up to a net misstatement of revenues of under $300 million. In 2005 Greenberg was out, and AIG lost the man who for almost four decades had served as a famously fanatical master of the details of the company's risk exposure.

As a result, perhaps, during the crisis of 2007 and 2008, AIG's risk exposure exploded, and the reinsurer of last resort turned out to be the U.S. Treasury, which made as much as $183 billion (the number dwindled over time) available for insured credit defaults. Though irritating to many observers, including both me and Goolsbee, the government rescue of AIG was not entirely inappropriate, inasmuch as the credit default swaps in question gave insurance to securitized tranches of subprime mortgages that were cherished by that same government and vastly multiplied as a source of expanded home ownership.

The government's eventual multi-billion-dollar profit on the deal may influence Goolsbee's future Nobel nominations. In any case, the chairman of the president's council of economic advisors exhibited as little understanding of banking and financial crises as the most intemperate Tea Partier or Occupy Wall Street ranter (who lack the excuse of advanced economics degrees from Yale and MIT and a full professorship at the University of Chicago).

Financial crises are no more a product of evil machinations than are hurricanes. If you build your house with the wrong stuff in the wrong place, with the wrong algorithm, you may be hit. One of Goolsbee's Nobel laureates in the AIG financial products division was Gary Gorton, a Wharton professor of finance who wrote his doctoral dissertation on "Banking Panics" and who was hit so hard he had to repair to Yale. Gorton contributed the algorithmic models behind AIG's credit default swap business and assured his bosses that his system could survive "adverse economic situations." He now comments acerbically that to understand the ferocity of adverse economic situations in actual banking panics it helps to have lived through one.

In his incisive book *Misunderstanding Financial Crises* (2012), a sadder but wiser Gorton catalogues eighty-six episodes in sixty-nine countries between the late 1970s and 1995.[1] These events inflicted average losses of between 10 percent and 40 percent of GDP, with the outlier Thailand suffering a 97-percent loss as a proportion of GDP in the Asian crisis of 1997. Despite a long "quiet period" between 1936 and the savings and loan crisis of the 1980s, the U.S. has suffered scores of such events.

Financial disasters befell so many different regimes—Japan, Mexico, Chile, Canada, capitalist mainland Asia—that ascribing them to an evil group of guys in one innovative but decephalized global insurance company seems a tad unscientific for a leading economist. But other economists have offered little further illumination. As MIT's Charles Kindleberger, author of the classic *Manias, Panics, and Crashes*, observed in 1984,[2] much of the profession is empirically bankrupt because it is no longer taught economic history.

What was going on, it turned out, was not conventional disruption of macroeconomic circular flows, but an acute information mismatch between short-term low-entropy money, which banishes transactional surprise, and the long-term high-entropy quests for upside profit in the assets behind it.

In *Wealth and Poverty*, I summed up this kind of mismatch in more poetic terms:

> Perhaps the central secret of capitalist success is its ability to convert the search for security, embodied in savings, into the willingness to risk, embodied in enterprise. The financial markets [chiefly in banks of various kinds] enact this crucial alchemy, turning fear into growth, caution into creativity, timidity into entrepreneurship, and the desire to conserve into the drive to build and innovate.
>
> This is a key capitalist industry, processing the abundant raw materials of anxiety about the future into the rare assets of faith in it and provision of productive facilities for it. One role is performed chiefly by an elite of businessmen, the other by millions of savers, but the two impulses are indissolubly linked at the heart of capitalism; they are the systole and

diastole of the production of wealth and the circulation of income.[3]

A financial crisis is an acute coronary fibrillation in this vital process of intermediation between the mismatched impulses at the heart of the economy. The mismatch is inherent in all economies and is most acute when entrepreneurial risk-taking is most unbridled. Runaway enterprise expands the gap between audacious creators and the fearful savers who fund them. It makes people doubt the financial bridges between the two impulses of growth, in banks and other financial institutions.

Financial panics, therefore, are to some degree inevitable. In any capitalist economy, enough goes awry from time to time to provoke fears of savers about the safety of their money. Their bank accounts lose their character as news-less, zero-entropy stores of value. Their money is no longer opaque; rumors of untoward entropy fly. The depositors suspect the bank has made high-risk investments. There may be no funds behind those barred cages! It's a Ponzi scheme! A bubble is about to pop. The Chinese are going to withdraw their money! The result of all this information leaking out from the normally marmoreal surface of banking is a mass effort to get the money back—a bank run.

Such bank runs tend to happen near the peak of a boom when people suddenly imagine that everything is too good to be true. Below the opaque surface of the loch, monsters seem to lurk. Rather than revelations of a suppressed catastrophe, crises may be growth spasms. Gorton presented evidence that countries that undergo financial crises on average grow much faster than countries that do not. Thailand, for example, has grown twice as fast as India for decades, though India has had no financial crises (perhaps because most of the Indian bankers are on Wall Street) and Thailand is the world champion rough-rider of runs. Gorton also reports that democratic countries emerge from crises with less damage than do non-democratic countries, which suffer more problems of trust.

In the midst of financial panic, nothing is as it seems. Marked to market on the wisps of volatile sentiment and fogged with fear from panicky ratings agencies, vessels of wealth are shattered in a storm of rumors. The crisis of 2008 suddenly made transparent the values that had been artfully rendered opaque when packaged as low-entropy money and debt instruments. The world gathered pruriently around the factories where banks produce neatly packed sausages of monetary worth from the bloody and turbulent processes of a capitalist economy.

Money economies are based on a particular information structure. At the bottom—in the "physical layer," as defined by information theory—transparency rules, and information is dense and specific. Here debt is based on detailed information. The originator of a mortgage, auto loan, business line of credit, venture outlay, angel investment, or bank card debt knows the specifics of the purchase, the borrower's credit record, and the value and recoverability of the items acquired or other specified collateral. At this level, information is high but liquidity is low—retrieving the money entails messy and costly procedures such as foreclosure, bankruptcy, or seizure of property.

As we move up the economic pyramid, through financial intermediation by banks and other institutions, we lose specific information and gain liquidity. Unlike water, which becomes transparent as it liquefies, financial instruments become more opaque as they approach the liquidity of money. This is not a flaw of money and demand deposits; it is their enabling feature. The history of the money in your pocket—the ultimate sources of its value—is unknown to you. That's why money and deposits can be used anywhere for any legal (or even illegal) purpose. If you had to prove the value of your money, it would not be liquid. It would take hours or days to negotiate even a simple transaction.

Creating liquidity—making zero-entropy money out of high-entropy investments—is the job of financial intermediation. Banks' output is debt. They do not manufacture it out of thin air, as many critics of fractional reserve banking seem to believe. In fact, banks tend to be stingy. Behind every loan they extend is a lien. Palpable collateral backs every debt. By guaranteeing the value of the debt at par, banks remove information and entropy.

The defining characteristic of money and deposits is no surprises. With no-surprise money, an economy can sustain billions of transactions a day and entrepreneurs can plan many years in advance. In their important book *Invisible Wealth*, Arnold Kling, formerly an economist at the Federal Reserve and at Freddie Mac, and Nick Schulz of the American Enterprise Institute explain how this works with mortgages:

> Investment bankers funnel funds through brokers, using only summary statistics such as the borrower's credit score, the ratio of the loan amount to the appraised value (known as LTV), and the broker's historical performance with the funding agency. Investment bankers then pool loans together. Banks, mutual funds, and pension funds that buy the pools know only the general characteristics of the pool—the range of credit scores, the range of LTVs, and so on. These pools may be further carved up into "tranches," so that if loans start to default, some investors will take an immediate loss while others continue to receive full principal and interest. *At each step in this layering process, some of the detailed information about the underlying risk is ignored....* As an intermediary layer is added, while the amount of detailed risk information is going down, liquidity is going up.[4]

The result, write Kling and Schulz, is that "the ultimate borrower—in this instance the homebuyer—pays a much lower risk premium." That truth, though, is trivial compared with the real economic role of the banking process: to transform mortgages or other long-term loans full of sticky and frictional high-entropy information into money so smooth and devoid of information that it can be taken to the mall or the gas station or the illegal nanny service or PayPal. It is money, after all, that everyone wants for transactions, not information. It is money that fuels the transactions of the market economy.

The purpose of banks is to drive entropy out of debt in order to produce money. Transparent money is an oxymoron. Financial crises have their roots in this fundamental difference between the supply and demand sides of the economy. The demand side—the money side—is nearly devoid of information; the supply side is full of it. The distinction derives from the difference between business cycles and transaction megacycles, between time to market for a business idea and the velocity of money. Real investment, which is long-term and specific and high-entropy, is inexorably in conflict with consumption and transaction demands for liquidity, which are short-term and ideally zero-entropy.

Looking at capitalism as an information system, it is easy to see its fragility. People want currency—which comes from the Latin word *currens*, "running"—and they want it now, to buy a burger or a grande latte or an iPhone or a broadband connection or a tankful of gasoline or a car or a house. They want liquidity, which comes from their bank deposits, their checks, or their credit cards. They don't want anyone to ask questions beyond those needed to establish their identities. If they wish to avoid the identity issue, they use cash, which is completely devoid of information. Meanwhile the value of their money is created entirely through productive activities on the supply side. It is, contrary to what Paul Krugman will tell you, the supply of goods and services that creates all that demand.

On the supply side, however, schedules are radically different. The producers of these goods and related wealth have to develop skills and deploy capital over many years to make the items available. Time to market may be decades. That Apple iPhone is full of Qualcomm microchips manufactured in a wafer fabrication plant that takes at least five years to build and equip. The chips are the product of a multi-month, seven-hundred-step manufacturing process that has been designed, debugged, and tested by engineers over at least a four-year period. During all that time, little or nothing is "liquid." The capital is tied up in cement and steel and silicon and chemical systems and photolithography equipment and fiber optic lines and real estate and air freight and expensively trained engineers and executives in companies around the globe.

Banks and financial institutions bridge the enormous gap between the blind immediacy of the demand side and the deliberation, delay, and information density of the supply side; between transactional cycles and business cycles. Most of the time, they accomplish this prestidigitation marvelously well, but not always.

Taking in liquid deposits and issuing long-term bonds, banks profit from the spread in the maturity mismatch. In the 2007–2008 crisis, the bank liabilities—debt, deposits, commercial paper, prime broker balances, and repos (repurchase and sale agreements, a form of overnight deposits)—were short-term, liquid, and guaranteed at par. They promised no surprises and no entropy. The assets backing these liabilities and collateral were high-entropy, subject to market vicissitudes, the frailties of human enterprise, conflicts of intellectual property, and acts of God such as hurricanes, earthquakes, wars, public panics, and the ousting of Hank Greenberg.

All financial panics reflect the collapse of the bridges of trust between short-term liquid liabilities and the long-term assets that back them up as collateral. The opacity of money and near-money gives way to fearful glimpses of information that cast doubt on their liquidity. Money and checks incur surprises. Your hundred-dollar bill is scanned and spurned. Even your traveler's check is rejected. Your fifty-dollar bill is discounted. When complete transparency is demanded, liquidity freezes and the economy relapses into an unmediated system of barter.

As the Stanford economist Ronald McKinnon has shown long ago,[5] financial development and intermediation is central to economic advance in both developed and undeveloped countries. It is a process of removing information from transactional media and suffusing information into the investment process, creating knowledge and wealth. The polarization of the economy between zero-entropy money and high-entropy investments is a source of the voltage that drives growth. It gives both sides of the economy what they want: the savers get low entropy, the investors get surprises, up and down. The polarity between the two impulses of capitalism is the source of both economic vitality and financial fragility. Economies stagnate and seize up when most investments become low-entropy (changeless big-company commodities and projects guaranteed by government) and money becomes high-entropy, full of surprises of devaluation and illiquidity.

The issue is not whether government should act in a crisis. As Gorton shows, in the history of capitalism, there has scarcely been a case of bank runs or panics where government stood aside and allowed the failure of the financial system. When money and banking are exposed to the world, the authorities will always rush in to cover their nakedness. When the mismatches of timing between instant money and long-term loans are revealed in a crisis, the government always mitigates the crisis with bank holidays, moratoria, delays, and other measures to rectify the phase differences between saving and investing, fear and faith. Before the creation of the Federal Reserve in 1914, a clearinghouse system arose to enable collective action by banks. But even that system could not deal with the repeated bank runs of the period without the intervention of a deus ex machina like J. P. Morgan. Since the creation of the Fed (after Morgan got tired or jaded), the frequency of crises has diminished. Deposit insurance has rendered bank accounts more opaque and money more robust. The Fed has often succeeded in tempering crises and making time by reconciling the phase clashes of capitalist money and investment.

What was different in 2008 was that all the politicians enlisted in the panic. Efforts to sort out bad investments from solid values marked to mist during the crisis gave way to efforts to read the minds of high-entropy government officials capriciously saving one company and dashing another. The Fed itself was engulfed by this process. The production of money became politicized and high-entropy, and the horizons of the economy darkened. As a result, some 30 to 40 percent of the economic profits during the mid-2000s came not from productive investment and entrepreneurial risk but from the zero-sum games of speculative investment and fiduciary risk shuffling. Public debt, under Parkinson's Corollary, expanded to absorb and stultify all other means of finance.

The government began guaranteeing everything—mortgages, deposits, pensions, healthcare, industrial conglomerates, leviathan banks, solar plants, small business loans, waterfront property, corn prices, college tuitions, windmills, kitchen sinks—except for the value of its currency, which since the time of Alexander Hamilton has been its job. Meanwhile, innovation became focused on a confectionary froth of social networking and financial derivatives. Little attention was paid to the physical layer underneath. Physicists began earning vastly more in finance than in the physics of entrepreneurial creations.

Could it be that the fundamental cause of the crisis was that the monetary system, alone among the structures of capitalism, lacks a low-entropy physical layer?

Over the centuries of monetary history, the remedy for unstable money has always been gold. Critics who say the gold standard has been eclipsed by an information standard based on the Internet do not grasp the essence of information theory, which measures the information content of a message by its "news" (expressed in digital form as unexpected bits or entropy). It takes a low-entropy carrier to bear a high-entropy, newsworthy message.

The 130,000 metric tons of gold that has been mined in all of human history constitutes the supreme low-entropy carrier for the upside surprises of capitalism. Without guidance from gold, currency markets are subject to political high entropy. They resemble a communications system without a predictable carrier that enables the information to be distinguished from the noise in the line. Such free-floating markets lack any objective means to differentiate the news (change in economic conditions and prices) from the "white noise" of currency flux. The chief source of new financial wealth in such a system is the exploitation of the gyrations of monetary noise by entrepreneurs of ignorance.

Money is an expression of productive services rendered, but by definition it is distinct from the services that are the source of its value. Without a baseline of gold, entrepreneurship in the world economy degenerates into the manipulation of currencies for the interests of profiteers and government insiders. This is a pathology of capitalism comparable to the manipulation of law by corrupt and profiteering lawyers.

A remedy for crises of money is gold. It may not be feasible, under current circumstances, to implement a real gold standard. But anyone in the business of dealing with money understands the intrinsic monetary signal of the price of gold and knows that this signal cannot be defied with impunity. A prodigal government that destroys the information content of money and prices will necessarily prompt a flight to gold.

Steve Forbes often speaks of the role of currencies as a standard of value, a measuring stick of the worth of the goods and services in the economy.[6] "Floating the currency is like floating the clock," he argues. "Let's say you floated the hour, 60 minutes in an hour one day, 55 the next, 85 the next. You would soon have to have hedges to insure against changes in the measure of time, just to calculate your hours of work in 'real terms.' You would have runaway sales of 'hour insurance swaps.' Same thing when you don't know what the value of your currency is going to be."

Banks are necessarily trust companies, and when they undermine the public's trust through trysts with government cronies, they endanger the most crucial role of banking.

A gold-based measure of money can play the role in economics that the Constitution should play in law. A gold link allows conversion between currencies around the globe; the absence of a gold link invites flights from fiat money and sovereign debt around the globe. Gold forestalls the volatility and high-entropy gyrations that spring from fears of government manipulation and banking fraud. It thus can extend the horizons of the world economy.

What it cannot do is end the intrinsic scandal of banking. That scandal derives from the maturity mismatch at the heart of any capitalist system of long-term, high-entropy supply and instantaneous, low-entropy demand. You will hear about this scandal wherever libertarians gather. There is no money in the banks. Fractional-reserve banking is a fraud and a delusion. It allows unlimited production of money by the banks, which can issue money at will.

The late Murray Rothbard and other goldbugs who believed in the scandal of banking propagated these fears. They believed that the gold had to be in the banks themselves and available for delivery at all times. The banks' ownership of a gold mine of capitalist assets was not enough. The money had to be reified and liquid at the same time, simultaneously frozen and fungible. That can never be.

The scandal of money is at the root of our wealth.

The Fecklessness
of Efficiency

THEORIES OF THE CRASH and panic of 2008 abound. From a run-away money supply ginned by the Federal Reserve to nincompoop regulators asleep at the wheel, from reckless and predatory "banksters" to greedy and sleazy lobbyists, from binges of government spending to orgies of mortgage lending, from gargantuan prodigality at Fannie and Freddie to promiscuous leverage on Wall Street and lawyering on K Street, there are so many miscreants and mistakes that in the end the picture is as opaque as a triple-A mille-feuille mortgage security insured to the hilt and with no one on the hook.

The fact is, however, that the crash has a clear and identifiable cause. That cause is a prevailing set of economic ideas that can be summed up as capitalism without capitalists—capitalism dominated by financial hypertrophy rather than technological vision and innovation. As Andrew Redleaf and Richard Vigilante put it in *Panic: The Betrayal of Capitalism by Wall Street and Washington*, "The ideology of modern finance replaced the capitalist's appreciation for free markets as a context for human creativity with the worship of efficient markets as substitutes for that creativity."[1] The result was a divorce of entrepreneurial knowledge from economic power.

Peter Drucker had an epigram for that: The successful executive pursues not efficiency but effectiveness.[2] He seeks not to do things right but to do the right thing. Devoted to efficiency and financial intricacy, in response to government mandates and guarantees, the policy makers, bankers, and money managers of America attenuated the ownership of the nation's assets. They pried apart the sources of knowledge from the sources of funds. A regime of capitalism without owners destroyed the information content in U.S. markets, prices, and "products." Breaking every close and informed tie between themselves and their assets, investors diversified, hedged, and insured themselves until they had not the slightest idea of the content of their holdings.

At his Whitebox hedge funds in Minneapolis, Andrew Redleaf was one of the few who saw it all coming. In December 2006 he famously told the world, "Here is a flat out prediction for the New Year. Sometime in the next 12 to 18 months there is going to be a panic in credit markets.... [T]he driver in the credit market panic of 2007 or 2008 will be a sudden, profound, and pervasive loss of faith in the alchemy of structured [mortgage] finance as currently practiced...."[3]

Festooned with government guarantees and rating agency laurels, these securities were deemed the world's safest investments despite the lack of any published details about their contents and prospects.

Redleaf first got a clear idea that something was awry at a road show in 2002 for New Century Financial, which was hustling up potential new investors. A go-go mortgage originator, New Century was hot stuff at the time. It supplied the raw materials for mortgage-backed securities, which in an era of near-zero interest rates had become the chief way for banks to make money and savers at home and abroad to get a return.

Growing at a headlong pace, gaining share against all rivals, New Century was on its way to becoming the second-largest issuer of subprime mortgages. It was on track to sell $135 billion worth of them to Goldman

Sachs for packaging into triple-A securities. For their (and our) pains, New Century's executives would make some $40 million in bonuses over the following three years.

In its road show, New Century ascribed its ascent to special expertise at finding mortgage customers who were actually good prospects despite having been shunned by stodgy conventional banks like M&T, BB&T, and Wells Fargo. Sitting in the audience, Redleaf thought this claim was absurd. "These guys are going to go broke," he said to himself. When the company spokesman asserted that New Century's average loan-to-value ratio was 75 percent, Redleaf smelled a runaway pack of rats.

He asked the spokesman, "Are you averaging 75 because the loan-to-value of most loans in your book is between 70 and 80? Or are you averaging 75 because you have a bunch of 50s and 60s along with a bunch of 95s and 100s?" Mortgages with down payments below 5 percent (a 95-percent loan-to-value rate) were vulnerable to any downshift in housing prices.

Redleaf knew that such small down payments are the best predictor of defaults. But the spokesman didn't think the breakdown between people who owned substantial equity and those who didn't mattered. "The average was what counted," and New Century had it covered. The spokesman did go on to explain that 85 percent of their mortgages were "cash-out refis," meaning that the borrower was refinancing his house in order to get cash out of it.

His nostrils stinging from the murine stench, Redleaf saw back in 2002 the essential corruption of the system: "A new buyer putting down 20 percent and the old owner taking money out of his house are doing profoundly different things. One is becoming an owner; the other is weakening his ownership. One is buying in, the other is selling out." Almost all the New Century customers were selling out.

By the time Redleaf was catching on to New Century, "the United States was almost a decade into a massive campaign to replace strong owners with weak owners in the housing and mortgage markets." This campaign would continue while New Century's executives were indicted for fraud in 2007. The campaign would extend to stocks and other assets acquired by index funds that had no need for specific knowledge because they bought a sample of the entire market.

The prevailing theory of capitalism suffers from one central and disabling flaw: a profound distrust and incomprehension of capitalists. Capitalists are owners who understand intimately what they own. For the last decade, the chief endeavor of capital markets has been to weaken all the disciplines of ordinary ownership.

In the place of owners were CFOs, who play the role of economists for individual companies, attenuating the accountability and responsibility of effective ownership and focusing on the benefits of financial efficiency. What distinguishes capitalism from other forms of economic management is that the capitalist cannot guarantee his returns. The key principle of investment is that its outcome is determined by the voluntary responses of customers. If there is a government guarantee, it is not a capitalist venture and there is no real owner and no real profit. But to a typical CFO, securing a government guarantee was more desirable than opening a new market.

In Richard Posner's view, this failure of corporate management became a "crisis of capitalism." A federal appeals court judge and one of the leading legal scholars of our age, Posner has recently authored portentous tomes titled *The Failure of Capitalism* (2009) and *The Crisis of Capitalist Democracy* (2011),[4] which describe our financial predicament as a new *depression* that shakes the foundations of democratic capitalism. Observing the spectacle of U.S. bankers paying themselves some $3 trillion in six years while undermining the financial system, Posner concluded that the new "Depression is the result of normal business activity in a laissez-faire economic regime—more precisely, it is an event consistent with the normal operation of economic markets.... Bankers and consumers alike seem on the whole to have been acting in conformity with their rational self-interest."[5]

Posner does not blame the government. "First, were there no government regulation of the economy, there probably would still have been a depression...." Most conservative and Austrian-school economists are ascribing the runaway housing bubble to the Federal Reserve's long siege of near-zero interest rates under Alan Greenspan. Posner acknowledges that "low interest rates alone increase the demand for housing because housing is bought mainly with debt." But he dismisses the effect of Fed policy, pointing instead to an array of mostly private excesses— "aggressive marketing of mortgages, a widespread appetite for risk, a highly competitive, largely deregulated finance industry, and debt securitization."[6]

As if to verify Posner's narrative of rational response to incentives, America's banker supreme, Charles Prince, CEO of Citigroup, declared in 2007 that "as long as the music keeps playing," he was driven to "get up and keep on dancing." According to Posner, the members of this chorus line were authentic capitalists, forced to follow the dance steps and pirouettes of deregulated rivals.

Consumers likewise behaved according to capitalist rationality by accepting mortgage bargains at a time of rising housing prices. It all made perfect sense in capitalist terms—self-interested agents responding to market signals of opportunity. It was not the greed of bankers or consumers or the obtuseness of regulators. It was the flawed logic of capitalism itself.

In Posner's analysis, an army of mathematicians, quants, and computer nerds invaded Wall Street and developed sophisticated hedging algorithms, diversification schemes, and insurance techniques that ostensibly reduced risk to near zero. Sophisticated investors then set out to harvest without peril the difference between the cost of the insurance and the yield of the investments. As Posner maintains, the explosion of leverage over the last decade seems to make perfect sense in capitalist terms. With close to zero risk, bankers are entirely rational to maximize their borrowing; the banks with the most leverage would win the largest returns.

The goal was to take high-entropy investments—mortgages with high probability of default, needing fundamental analysis—and turn them into the almost-money of low-entropy, triple-A securities. Triple-A securities are supposed to be devoid of surprises, which is what makes them almost money. But the pseudo-triple-A mortgage securities created by the banks, though stripped of all useful information, were not stripped of the one big surprise that most of them were worthless.

The system failed not because it was rational, but because rational choice in the face of massive ignorance—whether attributable to folly or deceit—is meaningless. Capitalism depends not on the freedom to choose but on the free flow of information across a low-entropy carrier. Corrupt the carrier with noise, and capitalism collapses. And the great corrupter of any carrier, the great generator of destructive noise, is power. And in this case the powers assembled were immense.

On the wrong side of the trades that brought down the economy—that is, the buyers of the bad mortgages and the filigreed bonds and the tranched truffles—were most of the world's central banks, the World Bank, the International Monetary Fund, Fannie Mae and Freddie Mac, Citigroup, Merrill Lynch, Deutsche Bank: all established institutions with intimate governmental connections. These banks, along with their regulators, the global financial constabulary, the universities, the charities, the most sophisticated politicians, were unanimous in their enthusiasm for the idea that the magic of structured finance could transform pigsties into palatial triple-A housing buys.

On the other side, unanimous in their mistrust of this financial sleight of hand, were hedge-fund entrepreneurs and other private venturers, such as Redleaf, John Paulson, Mike Burry, Steve Eisman, Greg Lippman, and perhaps twenty others of importance. A world away from the world's central banks, these firms were typically in the control of either a single manager or a small group of partners—people usually all well known to their investors and holding a majority of their own net worth in the firm. The majority of these owner-managers are self-consciously defiant of the methodologies that reign at the giant investment banks such as Goldman Sachs and Citigroup. They claim to be able to do the one thing that modern financial market theory asserts to be impossible—through careful research about their holdings, they claim consistently to produce above-average investment profits while incurring below-average risk.

Beginning in late 2005, a few of these entrepreneurs, fewer than two dozen all told, began to accumulate short positions on the worst mortgage securities. Chronicled by Michael Lewis, this "big short" was a frantic signal sent to financial markets warning of the peril the great banks had created, for even then it was not too late to prevent the looming catastrophe.[7] One of the heartbreaks of the financial crisis is that the majority of the worst loans were not written until the very end. Had the "music stopped" in late 2005 or early 2006 when the small band of hedge funds first made their move, close to a trillion dollars of the worst mortgages would never have been written. The bumbling banks and feckless Fannies and Freddies could have inflicted much less damage to themselves and the nation.

The reason the crisis was allowed to career out of control was simply that the blind side had all the capital—thousands of times more than did the entrepreneurs betting against them. The insurgents' short signal failed to emerge clearly through the higher-powered governmental and big-bank noise until the spring of 2007. By that time, the banks—still bravely estimating their potential losses at a maximum of a few hundred million—were already hundreds of billions under water.

Nearly devoid of relevant knowledge as they bought up subprime concoctions, these huge banks remained fortified by their power: their privileged positions at the automatic teller machines and discount windows of the Federal Reserve and vast leverage covered by federal insurance and Fannie and Freddie. These banks also boasted the approval of the U.S. Treasury and enjoyed a certain moral stature as the administrators and enablers of politicized federal housing policies. Meanwhile, attempting to shield the public from the supposed risks of entrepreneurial finance, regulators made it impossible for managers of independent funds to amass remotely comparable resources. Although as big as the entrepreneurs could possibly make it, the big short was nowhere near big enough.

The most exalted central banking authority of all was the Group of Ten in the Basel II process. This financial ultra-elite develops the rules of safety for national banks. Basel II actually *mandated* the purchase of the very sovereign bonds and subprime mortgage securities at the heart of the crisis as part of banks' required reserves. It endorsed structured collateralized debt obligations (whereby the actual viability of the underlying assets was opaque); and specifically opposed the ownership of individual mortgages (where relevant information was readily available). It thus explicitly espoused the crippling divorce of knowledge from power, information from capital, and accountability from ownership that caused the debacle.

The goal of financial reform should be to end this divorce between knowledge and power and ensure that most capital flows to the people who can wield it best, the entrepreneurs. By contrast, the reform that was enacted, the Dodd-Frank Wall Street Reform and Consumer Protection Act, with 2,323 pages of bromides and legalese enhancing the power of regulators, moves in the opposite direction.[8] It increases the distance between the people in authority and the people with entrepreneurial knowledge. It concentrates yet more power in the leviathans and their Treasury sponsors and Federal Reserve supervisors, and it imposes hundreds of new regulations on entrepreneurial finance. Dodd-Frank sacrifices information on the altar of ignorant power.

Economics from the time of Adam Smith has been focused not on the question of who should have capital but on bestowing the right incentives on whoever happens to have it. The economy, in Smith's—and ultimately Posner's—view, is "a great machine," a Newtonian mechanism driven by self-interested optimizations as inexorable as the force of gravity. Smith believed that the invisible hand of spontaneous order would assure that the public interest is served by the collective operation of self-interest. In other words, established economics holds that capitalism is an incentive system rather than an information system.

The key issue in economics is not aligning incentives with some putative public good but aligning knowledge with power. Business investments have both a financial and an epistemic yield. Capitalism catalytically joins the two. Capitalist economies grow because they award wealth to its creators, who have already proved that they can increase it. Their tests yield knowledge because they are falsifiable; they can be exposed as wrong. Businesses are subject to failure.

If business hypotheses are shielded from falsification by political protections or governmental subsidies or bureaucratic mandates, they cannot yield knowledge, and in most cases they destroy it. The power of politicized companies grows and congeals, but their knowledge degenerates into ideology and public relations. BP fatuously postures as "Beyond Petroleum" just before a new global oil and gas boom, and GE turns from productive manufacturing to financial arbitrage just before a crash, meanwhile devoting its residual manufacturing assets to confectionary energy projects such as windmills and defective light bulbs. Self-justifying leviathans—from Harvard University and Archer Daniels Midland to Fannie Mae and Goldman Sachs—loom over the economic and social landscape, distorting price and opportunity signals, etiolating the real enterprises in their shadows.

Unlike an inexorable, Newtonian "great machine," the economy is not a closed system. It will not function without knowledge and transparency—information flowing freely across a low-entropy carrier. Capital belongs in informed hands. It must move away from the blind side into the precincts of entrepreneurial learning and feedback, where knowledge can align with power.

14

Regnorance

ON MARCH 3, 2009, Ben Bernanke, chairman of the Federal Reserve Board, discovered a "huge gap in the Regulatory System."

This was astonishing. How could any gap have remained in regulatory coverage after several decades of expanding the Federal Register by more than ten regulations per day and by hundreds of thousands of pages? But Bernanke was clear. "There was no oversight," he informed those unsleeping watchmen of the Senate Budget Committee, "of the financial products division."

The particular financial products division that so exercised the chairman, one had to admit, was no small thing. It was the financial products division of American International Group (AIG), whose $2.7 trillion of credit default swaps had earned them Austan Goolsbee's nomination for the Nobel Prize for evil. Built over thirty years into the world's largest financial and insurance conglomerate under the leadership of Hank Greenberg—a recipient of the Bronze Star, veteran of two wars, liberator of Dachau, and pioneer of reinsurance—AIG went astray after 2005, when Eliot Spitzer ejected Greenberg from his office.

Banishing the company's legendary owner-entrepreneur was presumably a big step forward in the annals of oversight, but it somehow did not suffice to get AIG's management in order. Spitzer's attention may have wandered. In any case, within the next three years, while no one was looking, the AIG financial products division wrote those trillions of dollars of credit default swaps. A form of insurance mandated by regulators from the Federal Reserve in Washington to the G7 in Basel, Switzerland, credit default swaps were designed to offset the risk of subprime-mortgage-based securities, making them worthy of triple-A certification for sale to your grandmother, your pension fund, or to German, Irish, Chinese, and even Icelandic banks.

As part of the government's regulatory full-court press for housing finance, spearheaded by Fannie and Freddie, these swaps had risen from some $4 trillion in 2003 to $62 trillion in 2008. Such numbers can mislead. While often compared with global economic output of $68 trillion, they do not resemble real claims on goods and services. They are more reasonably compared with the $25 trillion in U.S. life insurance policies currently in effect or the roughly $75 trillion in other insurance policies. Credit default swaps are contingent on certain events, and those events (securities defaults) are less frequent than, for example, the usual occasions for life insurance payouts.

Still, Bernanke would have had a point about his gap if, indeed, in the dense forest of federal, state, and international regulations, there were no special pages or dockets or functional powers or supervisory authorities or official examiners or on-site inspectors or compliance monitors concerned with an AIG financial products division emitting perhaps more paper than the Federal Reserve itself.

Perhaps the financial products division had eluded Bernanke's own regulatory eye at the Federal Reserve ("I know nothing about AIG," he acknowledged at the height of the crisis). But all of AIG, especially the financial products division, was supervised and pettifogged by federal, state, local, and global beadles galore, in fifty states and more than a hundred countries. You have to be a professor at Princeton or a supreme bureaucrat in Washington to imagine that any problem of modern government could be attributable to inadequacy of regulatory coverage anywhere in the developed world.

The problem with the financial products division was not that its regulators were absent or impotent. Its head, one Joseph Cassano, was heard during the crisis to complain stridently about *over*-regulation of his domain. The problem was that the regulators, like most regulators, lacked relevant information. They were experts on the politics of the situation, but had no real command of the intricacies of the businesses within their purviews or any stake in their operations. They were not perhaps as ignorant about the implications of credit default swaps as regulator-in-chief Ben Bernanke was about the imperial reach of the federal regulatory state. But it was a formidable gap all the same.

The columnist Mona Charen pointed out that "Democrats added 11,327 regulations to the Federal Register in the first three years of the Obama administration (and that was before the big drivers—Obamacare and Dodd/Frank really got going)." And the *Economist* noted that America "is being suffocated by excessive and badly written regulation," including "flaws in the confused, bloated law (Dodd-Frank) passed in the aftermath of America's financial crisis."

From the perspective of information theory, regulation is mainly an effort to replace knowledge with power. In general, the more regulation, the less information. The finance professor Frank Portnoy and the Pulitzer Prize–winning journalist Jesse Eisinger, supposed experts, described the situation in an amazingly ingenuous article in the *Atlantic*, "What's Inside America's Banks?"[1] The authors demonstrated that they have nary a clue:

> Accounting rules have proliferated, as banks, and the assets and liabilities they contain, have become more complex. Yet the rules have not kept pace with changes in the financial system. Clever bankers, aided by their lawyers and accountants, can find ways around the intentions of the regulations

while remaining within the letter of the law. What's more, because these rules have grown ever more detailed and law-yerly—while still failing to cover every possible circumstance—they have had the perverse effect of allowing banks to avoid giving investors the information needed to gauge the value and risk of a bank's portfolio. (That information is obscured by minutiae and legalese.)

You don't say.

Portnoy and Eisinger imagine that the banks can reveal "how much a bank might gain or lose based on worst-case scenarios—what would happen if housing prices drop by 30 percent, say, or the Spanish government defaults on its debt?" Or if a jetliner crashes into the World Trade Center or a meteor destroys the Hoover Dam? Or an earthquake and tsunami devastate the coast of Japan? Or a hurricane wipes out southern Long Island or Iran destroys Tel Aviv? It should be obvious that banks cannot possibly know the effects of worst-case scenarios and would be subject to endless litigation and SEC prosecution under fair disclosure laws and subpoenas from six or more other major federal bodies if their executives tried to divulge their particular nightmares or paranoid priorities to the public through inquiring reporters from the *Atlantic*.

The government cannot have any clear idea what complex corporations like AIG are doing, let alone what wreckage "worst cases" might inflict. Regulators do not know how to make AIG better at what it does, how to improve its efficiency and effectiveness as a global insurance company. If AIG as an assemblage of the world's leading experts on risk management has no way of predicting the future, how can government regulators help? So regulators impose an array of rules on them that have the effect of distracting the company from its corporate purposes and toward government purposes, making AIG less like an entrepreneurial enterprise and more like a government bureaucracy full of lawyers filling out forms for regulators that aren't read until after something goes wrong.

In the case of AIG, the regulatory state sucked the corporation into the quasi-governmental domains of Fannie Mae and Freddie Mac, focused less on the usual insurance markets than on the reinsurance of subprime mortgage securities. We know how that turned out.

The rules from outside AIG took the place of actual knowledge residing within the company about its operations and markets—knowledge that the government lacked. Information entropy behaved like physical energy, dissipating and becoming unavailable. In particular, the wealth of vital knowledge in Hank Greenberg's head became unavailable, thanks to Eliot Spitzer. Rather than learning more about customers, AIG aimed at pleasing regulators by helping Fannie and Freddie fund housing.

Regulations that should be a low-entropy carrier, facilitating the operations of business, become high-entropy interference, reorienting business toward politics. As Hayek wrote in *The Road to Serfdom*, government should provide the rules of the road, a low-entropy role, not be the backseat driver.[2] Government should be carrier, not content.

The network of rules encouraged corruption and undermined honesty in banking. "Enron violated the spirit of the law well before it ever violated the letter of the law," John Allison explains in *The Financial Crisis and the Free Market Cure*. "This rules-based system is also a terror for honest CEOs.... Since the rules are extraordinarily complex, it is impossible for a CEO to know whether the accounting reports are technically correct. A great deal of energy is spent on technical rules compliance, often at the expense of a rational economic evaluation or full disclosure of more important considerations."[3]

This pattern of ignorantly intrusive regulation showed up everywhere during the crisis. The most expensive of all the bank failures administered by the Federal Deposit Insurance Corporation was that of the Independent National Mortgage Corporation—IndyMac—which cost the FDIC some $11 billion plus extensive losses for uninsured depositors and other creditors. As the former Fed official Vern McKinley reports, the Office of Thrift Supervision gave IndyMac high ratings until shortly before it failed in 2008—with as many as forty bank examiners on site. McKinley's book, *Financing Failure: A Century of Bailouts*,[4] is filled with similar stories about Bear Stearns, Lehman Brothers, Washington Mutual, Wachovia, and all the rest.

"In theory, CEOs report to boards, who then report to shareholders," explains John Allison. "While that's true of most businesses, in the financial services industry we only quasi-report to boards, quasi-report to shareholders, and definitely report to regulators."[5] Just for starters, they report to the Fed, the SEC, the Office of the Controller of the Currency, the FDIC, the Commodity Futures Trading Commission, and now the egregiously unrestrained Consumer Financial Protection Bureau. And that's just at the federal level. Allison describes the pressure on BB&T "to install mathematical models like our larger competitors. Wachovia and Citigroup (which both failed) were touted by the regulators as having 'best practices' in risk management based on their mathematical models."[6]

The reality, undeniable by anyone who has studied the piles of documentary evidence, is that, throughout the epoch of systemic collapse, every large institution was thronged with examiners, overseers, supervisors, inspectors, monitors, compliance officers, and a menagerie of other regulatory constabulary. In every case, regulators confidently reported that all was well, "sound," "stable," or solvent until days or even minutes before their subjects announced their need for billions of dollars in bailouts to prevent catastrophe, not only for themselves but for the entire global financial system, at which point the regulators sent up the know-nothing cry for massive government intervention.

The completeness of regulatory coverage was no remedy for its epistemic futility. The regulators could have no more knowledge about the future of the regulated companies than could the executives of the companies themselves. Knowledge is about the past; entrepreneurship is about the future. Regulations are rules based on past experience. Regulators are political appointees responsive to their bosses and to the rules. Only entrepreneurial owners take their cues from the subtle signals on the crests of creation.

As Peter Drucker has explained, business inputs and costs are determined within the business, but outcomes are not. Businesses are dependent on information held by customers and investors, and both can change their minds in minutes. Balance sheets can wither, assets atrophy, and deposits evaporate without any perceptible changes in the accounts for days, weeks, or even—in cases like Enron's—years.

What is true about all companies is overwhelmingly true of banks when there are capricious, high-entropy changes in monetary policy and interest rates by the Federal Reserve. As Allison put it, "The market is based on experimentation." A government agency can destroy this foundation, he explained. "We stopped the learning process."[7]

With the regulators necessarily in the dark, government officials tried to hide their ignorance with Keynesian incantations from the Great Depression. In terms of information theory, they turned up the power to compensate for the lack of information. The most readily available lever of power was federal spending. Federal spending based on borrowing money from banks that is loaned to them at 0-percent interest from the Federal Reserve is the epitome of dumb money—devoid of information and deadly to the real assets of the nation. As the Hong Kong economist Charles Gave explains, any related increase in GDP comes at the expense of the nation's balance sheet:

> GDP being a flow based concept does not take into account the deterioration of the balance sheet of the country.... Only the servicing the debt, not its absolute value, is entered as an expense in the GDP calculations.... Rising public spending leads to lower growth, which leads to lower cost of mone.... [giving the politicians] the most marvelous free lunch the world has ever seen.... What is the meaning of a debt to GDP ratio if the GDP is government spending financed by debt?[8]

If an economy is an information system, government money is usually noise in the channel. Federal appropriations should convey information about available resources and serious long-term policy commitments. Instead the government dumped money on politically-favored institutions (large banks) while information shriveled. Whatever benefit the economic system allegedly received from the trillions of government dollars was outweighed by the iatrogenic panic that the spending induced. The government worsened the crisis by terrifying the public, signifying that the doctor was even more ignorant and panicky than the patient.

The failure of Lehman Brothers illustrates this problem, writes McKinley: "While the argument for intervention was all the adverse consequences ... of a Lehman bankruptcy, Lehman had absolutely no incentive to reduce those adverse consequences."[9] It was a classic case of moral hazard, the tendency of uninformed insurance to invite the conditions against which it indemnifies.

Bush's Treasury secretary, Henry Paulson—"I have always been a devout proponent of free markets"—constantly appealed to the demands of the "markets" as reasons for action: "markets want to hear ... markets would not be satisfied ... markets are expecting ... markets need ... markets demand." But as Vincent Reinhardt has observed, "Rather than forecasting underlying values, financial markets were predicting government intentions.... The private sector lost its incentive to pump capital into troubled firms and gained an incentive to pick among the winners and losers of the government intervention lottery."[10]

In other words, the markets were not accumulating economic knowledge. They were trying to predict the exercise of government power. They were trying to second-guess Paulson, who thought he was second-guessing the markets. The government was trying to figure out what the markets thought the government would think about what the markets were thinking. With no idea of what was going on, Paulson was impotent to improve the situation in any way. So he needlessly and repeatedly made it worse.

In the absence of both knowledge *and* effective power in the government, authorities ratcheted up the flow of money. Following the crude logic of tumbling dominoes and falling demand, they pursued the remedy, as prescribed by Dr. Paul Krugman, of "more funds." *More gauze.* One treasury spokesman, referring to TARP, admitted, "It's not based on any particular data point. We just wanted to choose a really large number." And an assistant secretary of the Treasury, Neel Kashkari, explained, "We've got to scare the shit out of the [Congressional] staff." As Tim Geithner commented in a rare moment of insight, "You cannot go out and talk big numbers with regard to the capital needs of banks without inviting a run.... You'll start a freaking panic."[11]

Shortly thereafter Geithner was out there promoting panic. Without bailouts, without TARP, without the Helping Families Save Their Homes

Act, without Dodd-Frank, without this or that surprising federal convulsion decided the day before, the country would plunge into a new Great Depression.

The proximate cause of the crash was high-entropy government—political leaders and bureaucrats abandoning their objective, non-surprising, low-entropy duties and issuing gouts of conflicting high-entropy signals that so filled all the channels of the economy with noise that no message of reason could get through. The response to every troubling event was to raise the power of the transmissions, increase the size of the bailouts, amplify the noise from the government, and suppress the signals from the economy. The situation did not improve until the noise of federal money and power subsided a little as the Feds ran out of funds and people stopped listening to the domino-theory ululations. In the economy as an information system, money divorced from knowledge—dumb money—clogs the conduits of enterprise.

All this governmental harassment is most damaging to executives in smaller but ascendant companies, who can hardly hear themselves think amid the regulatory noise. The writer-entrepreneur Neal Freeman describes the problem from the small business frontlines:

> In my new life I buy and (attempt to) resuscitate small manufacturing companies. They desperately need capital and a rational tax regime. But even more importantly, they need relief from regulation, which has become a kind of man-eating plant choking off growth. The big manufacturers have lobbyists who largely write the regulations and compliance officers who grease them. A small manufacturer has no lobbyist and one part-time, overworked, litigation-frenzied sap who got drafted as the compliance officer when he missed a staff meeting.[12]

As the regulatory economist Alexander Tabarrok writes, "The problem is that even if each regulation is good, the net effect of all regulations combined may be bad. A single pebble in a big stream does not do much, but throw enough pebbles and the stream of innovation is dammed.... Ten good regulations may sum to one bad policy."[13]

Of all the high-entropy regulations of recent years, perhaps the most damaging to innovation is the Sarbanes-Oxley Act, passed in the wake of the Enron debacle. Titled the Public Company Accounting Reform and Investor Protection Act of 2002, it aimed to improve corporate accounting and to bring it into conformity with Generally Accepted Accounting Principles (GAAP). By requiring that most members of a corporation's board of directors be "independent" (meaning, in practice, that they have no knowledge of the company), the act mired board meetings in accounting trivia. It required a bidding war for independent board members with specialized accounting skills. It forced executives to sign piles of documents, full of legal and accounting jargon too time-consuming to read, even if they wanted to, while continuing their actual jobs. In essence, the act forced everyone in authority in a U.S. company to constantly lie about his knowledge of accounting details.

Most destructively, Sarbanes-Oxley (or "Sarbox," as it's affectionately called) diverted American venture capital to large company deals rather than small company innovations. Dumping a pile of pettifogging rules on all public companies, the act brought initial public offerings for nearly all creative small companies to a halt. Not only did Sarbox represent a tax of some two million dollars on all companies going public, tripling the average cost of an IPO, it skewed companies toward otiose finance and accounting and away from innovation. A survey by the American Electronics Association found that companies with sales under $100 million were spending 2.6 percent of their revenues, and often more than 100 percent of their profits, on compliance with Sarbox rules. A 2006 study by the Committee on Capital Markets Regulation found that the percentage of all global IPOs executed on U.S. stock exchanges had plunged from 48 percent in the 1990s to 8 percent by the middle of the next decade.

Yet despite its onerous high-entropy detail, the legislation had no discernible effect on large companies. The piles of new paperwork intensified the financial crisis, largely a series of legal and accounting imbroglios. Fannie Mae and Freddie Mac committed their accounting blunders uninhibited by the redundant certifications and re-certifications Sarbox required. As Alan Reynolds observed, "Investors saw through Enron and WorldCom's exaggerated earnings and hidden debts long before accountants or federal regulators did."[14] The problems of U.S. business are simply not an accounting or procedural problem that can be addressed by regulators. They are substantive problems largely caused by high-entropy government, above all by regulatory excess. Superfluous focus on accounting and other procedural details is preventing U.S. industry from competing with rough-and-ready rivals in Asia that can build thousands of factories and skyscrapers while the United States dithers with environmental-impact litigation, re-computation, and backup of accounting reports.

Ronald J. Baker, the philosopher-CPA and industry visionary, points to the deeper issue:

> Approximately 70 percent of the average company's value *cannot* be explained by traditional GAAP financial statements. Adding more arcane and picayune rules to GAAP, or converging existing GAAP with international accounting standards, will not solve this problem. The accounting model is suffering from what philosophers call a *deteriorating paradigm*—it gets more and more complex to account for its lack of explanatory power....
>
> CEOs have to create the future, not relive the past, and the only way to do that is with a theory of the business, and to get outside of the four walls of their organizations and connect with external reality—where all value is created.[15]

Another name for dumb money and regulation is demand. The owner, on the crest of creation, is facing in a different direction than the government, which is earnestly studying the rearview mirror of demand. This information gap between the supply side and the demand side shaped the financial crisis.

The supply-side owner, caught up in the ongoing evolution of entrepreneurial knowledge, understands his protean predicament. Shaping the plans and projects and experiments of enterprise, he is creating the vessel that he steers, and he leads in the light of his own creation.

Government operates in the darkness of the demand side. The financial crisis of 2008 was an illustrative display of ineptitude across two administrations—responding to the confused clamor and machinations of special interests, the cumbrous movements of monetary aggregates, and the temptations of spectacle and theater. Seeking to ordain outcomes from the demand side, the blind side, without information and by dint of dumb money, politicians sought to control investment by brute spending. They attempted to steer the ship from the shore with no idea where it was going.

The regulatory blindness of the demand side is most evident in the fields of energy and the environment. How much energy the United States produces, and how clean the environment is, depend entirely on supply-side capabilities. All the relevant information is on the supply side. Physical entropy cannot be abated on the demand side. If a nation fails to produce fossil fuel energy in sufficient volumes, people routinely strip the land of trees to heat their homes and beat their mules. Government may demand solar energy or electrical hovercraft or carbon sequestration or windmills or biofuels or hydro-cells or geothermal plants. But even if the goals were desirable, it could achieve none of them without a profitable energy economy that can sustain growth while the new projects are nurtured.

The alternative energy projects that enjoy Washington's favor are all creatures of panic over putative climate change. The environmental movement is devoted to the famous "hockey stick" graph depicting long years of more or less steady temperatures followed by a sudden spike. The hockey stick is refuted, however, by a raft of data from both history and science attesting to the existence of a "medieval climate optimum," several degrees warmer than today, when Greenland was green, the cherry blossoms bloomed early in Kyoto, southern Chinese plants grew in Taiwan, people settled higher in the Argentine Andes, and human civilization flourished.

The discrediting of the hockey stick, however, did not cause the movement to take up a different sport. Instead it is deploying a new hockey team composed of nothing but goalies. Its latest strategy is apparently to stop all industrial and energy development, and to prevent the construction of any fossil-fuel plant regardless of its efficiency and environmental hygiene. A single environmental organization claims to have stopped 253 new plants in the last five years.

If Ben Bernanke was dismayed to discover a huge regulatory gap relating to credit default swaps, imagine the consternation in the Environmental Protection Agency when American entrepreneurs, led by the petroleum engineer George Mitchell, found and exploited a huge gap in the new hockey team's all-goalie defense. They solved the energy crisis from the supply side. Mitchell pursued none of the anointed "alternative fuels"—no solar, wind, biofuels, or geothermal. Yet in a high-entropy play, he potentially reduced the carbon emissions from power plants more than a hundredfold. Perfecting a technique of drilling for natural gas by horizontal hydraulic fracturing of hard bands of shale a mile below the water table, he helped make available potential trillions of BTUs of gas and oil here in the United States.

These "fracking" techniques ignited the natural gas boom of the last five years, creating hundreds of thousands of new jobs and billions of dollars of new profits for U.S. energy investors. Even President Obama noticed this feat. In his 2012 State of the Union address and then in campaign debates, he brazenly claimed that all the new gas and oil was the fruit of his administration's energy research, even though every official of the EPA opposed fracking at every opportunity.

As exciting as all the new oil and gas itself is, *National Review*'s Kevin Williamson reports in "The Truth About Fracking,"[16] that this new technology is "that rarest of commodities: a regulatory success story." While the EPA dithered, the Pennsylvania Department of Environmental Protection (DEP) rushed "to do it right before the feds make us do it wrong." It was low-entropy regulation, mostly at the state level, specifying levels of clean-up, but not the techniques used to achieve it—rules of the road rather than backseat driving from Washington.

The chief problem of fracking is what to do with the volumes of brackish contaminated wastewater that it produces. In the Marcellus shale area in Pennsylvania, on which much of the work focused in 2012, the industry learned powerful techniques for purifying and recycling the waste. The Pennsylvania DEP required that water discharged into streams or rivers must be potable—causing what Williamson reports as a "sea change" for the industry. Once the water is potable it is too valuable to discharge. Clean water becomes a new salable by-product of natural gas production.

Williamson sums up: "Cheap, relatively clean, ayatollah-free energy, enormous investments in real capital and infrastructure, thousands of new jobs for blue-collar workers and Ph.D.s alike, Americans engineering something other than financial derivatives—who could not love all that?"[17]

The answer, sad to say, is virtually the entire environmental movement, whose regulatory spearhead is the EPA in Washington. While the data all favor fracking, the politics in Washington, with the juggernaut of rich upper-class environmentalists in the lead, all favor suppression. Hockey-stick graphs and goalie-stick policies converge to oppose all energy progress.

The economic growth crucial for technological advance and environmental integrity depends on huge production of power. The most efficient and cleanest power source is nuclear, at around two cents per kilowatt-hour. Virtually all environmentalists oppose it. The next most efficient source is combined-cycle turbine power plants fueled by natural gas. One hundred times less emissive than coal and 150 percent as power-efficient, it combines the direct pressure of gas expanding as it burns, with additional heat conversion for steam power. Because these plants can solve the claimed "problem" of carbon dioxide emissions far more efficiently than solar or wind alternatives, goalie-stick environmentalists around the country doggedly oppose these clean gas technologies too.

Because all the information is on the supply side, the regulators nearly always get it wrong. In a high-entropy move in 2010, while the fracking revolution transformed the energy environment on the supply side, the regulators proposed new demand-side Maximum Achievable Control Technology rules. Applied to trace level pollutants in 178 new categories, these rules gave the EPA the authority to decide what technologies are achievable in eliminating air pollution. But the EPA had no relevant knowledge of what technologies are achievable or sustainable in a free economy, where new tools have to be tested under fire and need to attract new investment. Substituting power for knowledge, the EPA imposed high-entropy rules on industrial boilers that are totally unachievable.

"The EPA is happy to let you operate a factory in the United States today as long as your emissions are zero," observed one covertly sophisticated state regulator to chemist Richard Trzupek, who described the program in his book *Regulators Gone Wild*.[18] Since American manufacturers cannot win as long as they continue to operate in the U.S., the result of this demand-side approach is removal of manufacturing to China, where emissions of mercury and other pollutants, as well as of carbon dioxide, are nearly five times as high per thousand dollars of national output.

The Obama Administration deferred application of some of its more extreme environmental measures until the year 2014, apparently in the belief that the economic damage will also be delayed. But prospective changes in demand-side regulations, which cannot achieve their goals, immediately change the information environment for all manufacturing assets of the United States. Investors in 2013 summarily devalue all projects that come to fruition after 2014. Devalued assets lead to declines in industrial innovation and resulting capital losses and real estate depreciation.

With new manufacturing outlays stifled and initial public offering stymied by Sarbanes-Oxley, investment shifts to finance, insurance, real estate, and mandated topiary greens. Regulatory overreach leads to hypertrophy of finance and litigation, and even transforms venture capital from a force of profitable creativity into a lobby for subsidies and government mandates that burden the national economy rather than stimulate it.

Thus the high-entropy demand for specific technologies distorts and depletes the entire information fabric of the system. As a demand-side force, regulation must be low-entropy and predictable because businesses on the supply side can respond to new requirements only over long spans of time. While the EPA can promulgate new requirements overnight, reverse them the next week, and change their focus in the third week, the companies that must comply with them depend on investments and developments that take many years to achieve.[19] The EPA's new rules deplete the very capabilities needed to comply with them. They devalue the assets that are indispensable to innovation.

High-entropy regulation destroys the environment that it seeks to enhance. As Alexis de Tocqueville wrote 175 years ago of guardian government, "It covers the surface of society with a network of small complicated rules, minute and uniform, through which the original minds and the most energetic characters cannot penetrate, to rise above the crowd."[20] High-entropy regulation crowds out high-entropy enterprise. Whether in finance or manufacturing, environmental protection or innovative investment, the volume of rules becomes noise in the channel.

California
Debauch

THE DESTRUCTIVE CHANGES in the U.S. economy stem from a radical shift in government policy, transforming the entire environment of information and law under which venture capital functions. Leading the country on this retrogressive course is California, the state that has served as the vital center of venture capital and growth in the United States ever since Hewlett Packard opened up for the sale of oscilloscopes in a garage in Palo Alto.

A headline from the *San Diego Union-Tribune* of January 6, 2013, tells the story: "Wealthy Feel Pinch of Trio of Tax Hikes." The story refers to a new 39.6-percent federal tax rate on high incomes, a 3.8-percent Obamacare surtax, and the new 13.3-percent income tax on successful Californians. Robert Shillman, the chairman of Cognex Corporation, a leader in robotic imaging, remarked, "The increase in the already-high tax rate is a strong disincentive for people who live and work in California. I have friends who already have left for Florida and for other states that either don't have the personal income tax or that have ones that are far lower than California's."

Then he added: "The reason this is happening is very clear: There are more 'takers' than 'makers' in our society, and this is leading directly to the decline in free enterprise and capitalism, and this will inevitably lead to the decline in the standard of living…not only in California, but also in the entire country."[1]

California's new taxes on entrepreneurs in 2012 got most of the media attention, but the state had turned decisively against growth and enterprise two years earlier. In what may have been the most consequential election of 2010, Californians voted against the repeal of Assembly Bill 32, the Global Warming Solutions Act. Passed in 2006, AB 32 ordained that the state's emissions of so-called greenhouse gases be reduced to 1990 levels by 2020, a 30-percent drop that would require the depopulation of the state. An 80-percent reduction is mandated by 2050—a level that, if accomplished globally, would bring the carbon dioxide content of the atmosphere down to ice-age levels, below which plant life shuts down and human beings would have to stop breathing. The effort to cap all energy production in California, together with an unsustainable $500 billion in public pension liabilities and rapidly swelling budget deficits— exacerbated by driving business taxpayers out of the state—presents a stark picture of a state in economic dementia.

The voters of the Golden State have somehow failed to notice that the advance in conventional "non-renewable" energy over the last century—from wood to coal to oil to natural gas and nuclear—has already produced at least a 60-percent drop in carbon emissions per watt. In the words of the natural gas pioneer Robert Hefner, "As man travels down the energy path from solid wood and coal to liquid gasoline and to gaseous natural gas and hydrogen, the progression is one of carbon heavy to carbon light; from complex chemical structure to simple; from toxic particulate emissions to no particulate emissions; and finally, from high CO_2 emissions to no CO_2 emissions."[2] Thus the long-term California targets might well be achieved globally in the normal course of technological advance. Unlike the holocaust of ingenuity and money, moreover, an organic advance of energy efficiencies will readily spread around the world without mandates and subsidies.

The obvious next step in this energy evolution is the aggressive extraction through fracking of the several trillions of cubic feet of low-carbon natural gas that have been found in the United States over the last two years. These discoveries have ended the "energy crisis" in the United States in material terms and made it merely a problem of regulatory overreach and political nihilism.

The California vote, however, gives a patina of public support to an economy-crushing drive to suppress carbon dioxide in natural gas and everything else. Already mounting a scare campaign against the new natural gas extraction methods, California Democrats apparently expect to pass on the huge costs of their policies to the rest of the country by way of the federal government—a reverse gold rush.

Masking the bailouts for the state will be subsidies for green jobs and stimuli for politically motivated research and development, so-called "feed-in tariffs" for nuisance energy suppliers, and a "smart grid" boondoggle that can adapt to dumb and erratic power but is more complex and vulnerable to sabotage. Eventually there will have to be a costly environmental cleanup for the toxic wreckage of windmills and solar panels (replete with lead and cadmium). When all these efforts to save California fail, the last resort will be a debt-withering siege of inflation that depreciates much of the nation's remaining wealth.

In a parody of the supply-side economics of creative destruction, advocates of AB 32 envisaged "alternative" energy sources creating new jobs and industries and replacing existing fuels. Thomas Friedman's *Hot, Flat, and Crowded*[3] is the bible of this delusional sect, which has captured much of Silicon Valley. This economic model sees new wealth emerge from dismantling the existing energy economy and replacing it with a medieval system of windmills and druidical sun temples. But the destruction of the workable and efficient energy system we have does nothing to enable a new one.

America's "greens," led by Amory Lovins and Al Gore, offer the indignant retort that non-renewable resources, such as oil and gas and nuclear energy, receive far more government subsidies—some $73 billion in 2008—than the $29 billion spent on the new green alternatives. It is instructive to look deeper into this claim. Much of the so-called subsidies for conventional energy are actually deductions for tax payments these companies made overseas. It is important to maintain the distinction between a recipient of a subsidy and a taxpayer. The conventional energy companies, unlike the renewable ones, make vast net tax contributions while supplying 98 percent of the nation's energy.

Even accepting the nonsensical claim that tax deductions and depletion allowances are a governmental handout for "dirty fuels," the so-called direct subsidies or tax benefits per watt are roughly twenty times greater for the renewable fuels. Meanwhile the carbon dioxide suppression caps and mandates represent a confiscatory new tax on all conventional energy sources, while all the accumulating environmental litigation and legislation pose an insuperable barrier to innovative new nuclear facilities.

The chief so-called renewable energies that the greens believe can replace oil, gas, and coal are wind, solar, and biomass. More energy is consumed in their production, transportation, and adaptation to the power grid than they yield in new power. Windmills, for example, not only deface the landscape but are too erratic to supply reliable base power and thus require a non-renewable backup. The venture investor and engineer Andy Kessler has calculated that because of the inefficiency of alternative sources, every dollar of new wages for green workers will result in at least a dollar and a half of reduced pay and employment for California's other workers. The damage, sadly, will not be confined to the state but will also displace and impoverish workers across the country.

During the career of "Climate Chancellor" Angela Merkel, Germany has become the cynosure of alternative energy movements, with forty-seven national imitators so far, including the United States. Subsidizing solar at double the U.S. rate for twenty-five years, Germany has so far failed to produce a discernible increase in the solar share of total energy use, still around 2 percent, while wasting some $70 billion in the process. To supply all their electricity with solar would cost some $3.5 trillion.

Unrecognized, however, is the most punishing effect of the California legislation. It has debauched and stultified America's and California's most important asset: the information processing and learning curve of its venture capital industry that accounts for the nation's technological leadership, military power, and roughly a fifth of GDP.

Venture capitalists are America's most valuable financiers because they fund the most creative entrepreneurial experiments. Rather than contriving intricate money-shuffling devices like many on Wall Street, they have funded the most important breakthroughs in global technology, from Intel, the world's leading microchip company, to Genentech, the revolutionary biotech firm that sparked a new industry in health care. I myself have devoted much of my career to chronicling the venturers' triumphs, and today I devote much of my time to my own more modest venture capital pursuits. I have discovered that the new Silicon Valley is Israel, which continues to focus on fundamental new technologies rather than on green frivolities.

The key to the green victory in California's election was the massive support of the state's high-tech industry. Against the feckless outlays of a few oil companies, totaling some $10 million, the greens put up $31 million, most of it contributed by venture capitalists. Under Al Gore's influence, millions poured into the green campaign from such venturers as John Doerr of Kleiner Perkins Caufield & Byers and his former colleague Vinod Khosla, Eric Schmidt and Sergey Brin of Google, the legendary Gordon Moore and Andrew Grove of Intel, John Morgridge of Cisco, and John Fisher of Draper Fisher Jurvetson, along with a passel of descendants of the late David Packard. The campaign even managed to shake down a contribution from the state's public utility, Pacific Gas and Electric. Those hapless Texas oilmen from Valero and Tesoro did not know what had hit them.

I am not so cynical as to imagine that my venture heroes would spend $31 million to save their sickly green portfolios if they did not think they were doing great good in the process. Most of the donors seem to have suffered midlife befuddlement brought on by plethoric wealth, and have quixotically set out to save the planet at the expense of the U.S. economy.

The green gaggle of venture geniuses in California are simply out of their depth in a field they do not understand. All these venturers made their fortunes in digital information technologies. As Eugene Fitzgerald, a professor at MIT and a microchip and solar innovator, writes in his tract *Inside Real Innovation*, "Early stage investors hoped that energy would be the same as the semiconductor-PC-information technology sector but alas...there will be no new form of energy analogous to the birth of the transistor or integrated circuit...that offers very dramatic performance gains...without breaking the laws of thermodynamics."[4]

The mistake of the greens in hoping for a fundamentally new form of energy is that they are working the physics, not the information. They are not transcending physics with information, they are trying to defy the laws of physics in a physical context, trying to beat physics on its own turf. In other words, they are attempting to create perpetual-motion machines. Even if there were an energy crisis, and energy were to become as scarce in the world as water is scarce in Israel, then, as in Israel, the response would be informational—to use information technology and information theory to extract and apply the valuable energy in the most powerful and efficient way.

Solar panels are not digital. They may be made of silicon but they benefit from no magic of miniaturization like the Moore's Law multiplication of transistors on microchips. Setting the size of photoreceptors are the wavelengths of sunlight, which are microns wide, a thousand times bigger than the nanometers of new chips. Biofuels are even less promising. Even if all Americans stopped eating (about a hundred thermal watts per capita on average) and devoted all their current farmland to biofuels, the output could not fill more than 10 or 20 percent of our energy needs (more than 11,000 watts per capita on average). The energy expert Bill Tucker (who predicted the natural gas abundance back in 1978 in an article in *Harper's*) explains that virtually all the energy in the atom resides in its nucleus.[5] Twiddling valence band electrons in solar cells can yield only an infinitesimal amount of energy compared with nuclear and chemical technologies, and because the renewable fuels are inefficient, they are horribly wasteful of the precious resource of the arable surface of the earth.

Matt Ridley makes the argument cogently in his landmark book *The Rational Optimist*: "Roughly 5 percent of the world's crop land has been taken out of growing food and put into growing fuel (20 percent in the United States).... Not even Jonathan Swift would dare to write a satire in which politicians argued that...it would somehow be good for the planet...to give up food-crop land to grow biofuels...thus driving up the price of food for the poor [who] spend 70 percent of their incomes on food. In effect, American car drivers were taking carbohydrates out of the mouths of the poor to fill their tanks."[6] Nonetheless, proving that no policy is too perverse or foolish for the environmentalists of the Obama administration, in 2012 the United States hiked the proportion of ethanol in gas from 10 percent to 15 percent.

Following Peter Huber in *Hard Green*,[7] Ridley makes the key point: "A sustainable future for nine billion people on one planet is going to come from using as little land as possible for each of people's needs."[8] From wind to solar to biofuels, renewable resources waste land, capital, and labor, and thus are an environmental catastrophe.

The general Gadarene rush for green subsidies in Silicon Valley thus directly destroys much of what it claims to help. But its effect on venture capital may be its least recognized problem, for most people do not grasp the centrality of venture capital to the U.S. economy. All the key technologies that sustain American wealth and power were developed by companies crucially supported by venture capitalists in California. None was more important than Kleiner Perkins Caufield & Byers, the firm led by Doerr and Khosla, which funded scores of vital ventures, including Apple (the leading innovator in consumer electronics), Applied Materials (the paramount producer of wafer fab capital gear), and Sun (once the leading innovator in workstations, now the hardware arm of Oracle, also supported by venturers). Other Kleiner Perkins beneficiaries are Amazon (the pioneering web retailer), Google (the spearhead of the emerging digital economy), and an array of microchip and fiber-optical investments that sustain the Internet and military innovation.

Kleiner Perkins has now descended to funding such politically dependent firms as MiaSole, Bloom Energy, Amyris Biofuels, UpWind Solutions, and Segway. Many of these ventures wield formidable technology and employ thousands of ingenious engineers, but they are mostly wasted on the photovoltaic conversion of sunlight into ambient pork and pelf. Other venturers were lured into funding the notorious solar panel manufacturer Solyndra, which received some $500 million in federal subsidies and a campaign visit from Barack Obama before slipping into bankruptcy in late 2011 (as I and many others had predicted in 2010). Many of these green companies, behaving like the public service unions they resemble, diverted some of their government subsidies into the AB 32 campaign for more subsidies.

What is the remedy for this disaster, which has transformed venture capitalists from heroic contributors to American innovation, capital formation, and government revenues into a pack of grubby petitioners for pork? Ultimately the movement must be stopped by a change in policy, a transformation of the infrastructure of laws and subsidies, mandates and media pressures, within which the venture capitalists operate. A powerful array of legal and political forces is mobilizing to suppress U.S. energy production in the name of the spurious man-made global-warming crisis. These forces of informational distortion will ultimately consign the U.S. to second-class status in the world economy regardless of what happens to the federal deficit.

The United States cannot undertake a costly, inefficient, and disruptive transformation of the energy economy, estimated to cost some $45 trillion over forty years, while meeting our global military challenges and addressing the debt crisis. The facts are binary and intractable. Alternative fuels cannot begin to meet American economic and military needs. Continuation down this path means the end of U.S. world leadership. Mandates, caps, and subsidies to downgrade our infrastructure will not create enduring wealth or jobs. But they deplete scarce and precious technological and entrepreneurial resources.

Closing the deficit will do nothing for the United States if we give up our technical prowess. With hard technologies in eclipse and thousands of engineers diverted into efforts to build perpetual-motion machines, even venturers who shun energy investments focus on such trivialities as social networking schemes. But Zynga, Farmville, Facebook, and Angry Birds will not sustain the U.S. economy, let alone our defense. Kleiner

Perkins's great new hopes are Twitter and Groupon. Although some social networking companies will aid the economy, the focus of venture capital on such distractions is an ominous sign for the future of U.S. world leadership.

The wealth and technology brought by economic growth are indispensable to a clean environment. And the chief danger to growth is a collapse of U.S. and global innovation. Mandates for a new and inferior energy system will retard the technological progress that could reduce real pollution and other environmental damage around the globe.

The environmental movement has palsied two generations of American youth. It has diverted much of our high school curricula into the phony field of environmental science. (As legendary physicist Richard Feynman observed, "If a science has an adjective it probably isn't science.") At the same time, the movement has turned many universities into apocalyptic nature cults that divert money from education to an obscurantist debauch. Seventy-two percent of Harvard students in late 2012 actually voted to have their university disinvest from all fossil fuels. This movement has already corrupted most branches of government with a carbon dioxide fetish. Now it is debilitating America's most precious venture assets.

The green mandates, subsidies, and intellectual confusions inflict damage far exceeding any problem of federal debt. Directly targeting the supply-side of the economy, they erode the value of every entrepreneurial initiative in the U.S. economy.

16

Doing Banking Right

DON'T SOLVE PROBLEMS.

When you solve problems, you end up fueling your failures, starving your strengths, and achieving costly mediocrity.

You end up propping up the past in the name of progress.

Pursue opportunities is the counsel of the capitalist.

Deep in the recession of 1982, with interest rates spiking above 20 percent, inflation in double digits, unemployment near 10 percent, and America's financial institutions in a dire and portentous crisis, Bob Wilmers resolved to go to Buffalo, a dying city at the core of Rust Belt decay, to take over a failing bank.

In the midst of our post-millennial depression, many Americans labor under the delusion that the debt and banking crisis of 2008 was unprecedented since the 1930s. This is a voguish vanity. The crisis of the 1970s and early 1980s following the end of Bretton Woods gold convertibility in 1972 was at least equally dangerous, was managed with comparable ineptitude by Republican and Democratic administrations, was accompanied by similar monetary excesses of sub-zero *real* interest rates (adjusted for inflation), and precipitated a comparable cascade of banking failures.

The former chairman of the FDIC William Isaac summed up the 1970s episode:

> A two-and-a-half year recession with unemployment reaching 11 percent, the massive insolvency of the thrift industry, the bursting of the bubble in the energy sector, a depression in the agricultural sector, a rolling real estate recession that wiped out major regional banks throughout the country, the failure of some 3000 banks and thrifts, and the insolvency on a market value basis of virtually all of the country's largest banks because of massive investments in real estate and Third World debt.[1]

It took remarkable vision and resourcefulness to see in this imbroglio a supreme opportunity. On the national level, President Ronald Reagan did, sizing it up and seizing it. Following his own contrarian vision, he accepted an acute and wrenching recession. Joining Paul Volcker's monetary garotte with across-the-board tax-rate reductions, Reagan chose policies seen by most economists to be at war with themselves and thus self-defeating: "tight" monetary policy and "loose" fiscal policy. In the process he established the credentials of supply-side economics and forged a historic record, launching a twenty-year boom.

In the banking industry, no one acted more boldly and efficaciously than Robert G. Wilmers. Although a liberal Republican whose political baptism came as deputy finance commissioner in the administration of John Lindsay in New York City, Wilmers provided an example of business leadership in the flinty contrarian spirit of Reagan. What Wilmers accomplished and what he learned belies much of the usual analysis of the recent crisis as a profound failure of capitalism or an inexorable result of uncontrollable events.

Sandy-haired, educated at Exeter and Harvard, and married to an elegant blonde art historian from the house of Crowninshield, Wilmers found himself at forty-six a vice president at the House of Morgan. After a rewarding three years running Morgan's three-hundred-employee private banking operation in Brussels, he was unenthusiastic about returning to the New York home office for a wealth-management gig. His azure eyes began to wander.

After a lucrative turnaround of Multibank in Quincy, Massachusetts, Wilmers saw that bank shares in 1982 were dirt cheap. The financial turbulence was intensifying. A sovereign debt crisis like the one that has engulfed Europe was erupting in Mexico and points south, and resonating fatefully in New York money centers. First Pennsylvania, the largest bank in that state, fell insolvent in 1980 when its portfolio of long-term fixed-rate government bonds sank in value as interest rates soared. Penn Square Bank, a high-flying national bank in the oil patch of Oklahoma, crashed in July 1982. Reverberations spread through larger institutions such as Chase Manhattan and Continental Illinois, for which, in a precursor of 2008, Penn Square was servicing doubtful real estate loans. The Butcher banks of middle America were collapsing. Most of the nation's S&Ls and other thrifts were in jeopardy and headed for losses as sizable as a share of GDP as the collapse of 2008.

Intriguing Wilmers and his partner George Perera was the Buffalo institution M&T—owned on Wall Street by a holding company called First Empire State—which had been founded in 1856 as Merchants & Traders Bank to serve the workers on the Erie Canal. In 1981, M&T was selling for one-third of book value, meaning it might be most profitable to auction the assets and bolt, which is what cynics in Buffalo imagined the Wilmers team had in mind. After forty years under non-owner management, M&T " was not on the fast track," Wilmers comments, "and with no concentration of ownership, there were vultures hovering in the air. At some point the previous management just got old."

Exacerbating the plight of this bank and many others was mark-to-market accounting, which mandates the assumption in the balance sheet that all assets, including long-term bonds, will be sold off at the price prevailing in the market at the time—even if, as is often the case in a crisis, there is no effective market. (Imagine that whenever a nearby house was sold at a distress price, you got a margin call from your bank demanding repayment of your home equity loan).

First Empire, owner of M&T, had nearly two billion dollars in assets (small for a bank) but negative net worth. It had 2.8 million shares on sale for upwards of ten dollars a share (equivalent to fifty cents today after splits). Wilmers and his partner ended up with 23.9 percent of the bank and two seats on its board. M&T's key attractions were direct flights to and from New York City (Wilmers planned an active role on site) and ownership of some securities with short maturities, which could give him a little time to maneuver. No one exulted. No one supposed that this small institution would eventually make Wilmers perhaps America's most successful banker and a paladin in finance for Warren Buffett.

The crucial aspect of Wilmers's venture was his willingness to take active ownership of the bank. He did not subsume it in a distant holding company and collect dividends. He did not transform it into a heavy originator of mortgages and other debt to be passed on to distant owners. He did not diversify it with holdings in other banks or unrelated ventures.

Wilmers moved into a Buffalo condominium in an august but jaded gothic revival tower at 800 West Ferry Street (where, so said the ad, "classic Buffalo grandeur meets modern living"). He showed up in his Buffalo office at least three days a week for thirty years and devoted most of the rest of his time to building the bank up and down the Northeastern seaboard. "I was a lone ranger," he acknowledged, "and I saw an opportunity."

At his first board meeting in November 1982, however, Wilmers discovered the downside of his investment. The bank was cheap for a reason; its asset pool mixed wealth with toxic waste, like a compound of the city's most famous sites: the Erie Canal and the Love Canal. During these days of real estate doldrums and Latin American debt, it financed a lot of real estate projects and held a lot of Brazilian and Mexican paper.

Then he encountered the dung on the downside. Six percent of the company was a rotten-borough subsidiary called the First National Bank of the Highlands, near Poughkeepsie. This operation turned out to have been serving as a financial front for organized crime, moving up into the

Hudson Valley from Florida and New Jersey. Wilmers's predecessor Claude Shecter sheepishly reported the bad news, which required a restatement of earnings, a delousing of the loan portfolio, and the delay of Wilmers's first annual meeting.

The "tip of the iceberg," reported Shecter, was that six unrepaid auto loans had been issued to known gang leaders. When he had dispatched a junior banking officer to inquire about one of the loans, the recipient took a key out of his pocket and scratched a long line down the side of the banker's car, warning, "if you bother me again this will happen to you." When Wilmers assigned an executive from Buffalo to run Highland, the local Mafia had his phone line tapped. All in all it was not a promising start for the Ivy-educated New Yorker and his upstate venture in hands-on capitalism.

During Wilmers's thirty years in Buffalo, the city lost 10 percent of its population and became "older, sicker and poorer." Wilmers, however, engaged deeply in the community and found the best opportunities there. He followed a conservative lending strategy based on intimate knowledge of his borrowers. He expanded the bank's branches and invested heavily in ATMs. He also made a non-dilutive acquisition every year, applying his experience in Buffalo to other Eastern markets, from Westchester County to Harrisburg, Pennsylvania.

For his first ten years, the bank grew more than 20 percent per year and attracted the attention of someone Wilmers's secretary described as "a man in a rumpled suit named Buffett." Already a billionaire, Buffett owned the *Buffalo News* and had a token holding of one hundred shares of M&T, which had been rising apace with the shares of Berkshire Hathaway. Buffett told Wilmers that he would like to do something together with him.

In 1991, they got their chance when the old Buffalo Savings Bank was put up for sale. The bank, headquartered in a sensational Beaux Arts gilded rotunda in the heart of Buffalo, had been recently renamed Goldome in the midst of a decade of growth across the state. Goldome was the largest savings bank in New York State and too big to swallow without help. Wilmers called Buffett and asked him whether he was serious about a joint venture. Buffett agreed to do whatever was needed to enable the acquisition, and he ended up investing some fifty million dollars in shares of M&T at around ten dollars a share.

Over thirty-four years under Wilmers, the shares of M&T appreciated nearly as fast as Berkshire Hathaway's, at a compounded annual rate of 20.6 percent per year including reinvested dividends. In 2006, the shares reached $120, up 240-fold since Wilmers bought the bank and twenty-four-fold since Buffett invested in it. The key to the growth was steady and prudent organic expansion of customers and deposits and twenty-three successful acquisitions in twenty-three years.

Between 2000 and 2010, M&T's growth per share was highest among the top twenty U.S. banks. For 140 quarters through 2011, the bank has been profitable. Alone among leading banks, M&T went through the banking crisis without issuing dilutive shares of stock or missing a dividend.

Meanwhile, its competitors, following the "dance while the music plays" strategy, fell catastrophically back, with 2,653 outright bank failures, including 140 in 2009 alone, when the banking industry as a whole—outside of the five bank-holding companies with the largest trading revenue—lost $7.4 billion.

As Wilmers explained, "Our performance has hinged on what we call our conservative credit culture. That means that we realize that the only good loan is one that will be paid back. And, yes, that means that during the recent years of irrational exuberance in the financial industry we did a pretty good job of staying away from the punchbowl." M&T's charge-offs as a percentage of loans in 2010 were lowest amongst all its large regional and super-regional peers, and only about one-quarter their percentage. Meanwhile M&T ranked sixth in the nation among all banks in small business loans.

Eloquent, knowledgeable, and respected, Wilmers has emerged as a high-profile critic of the supposed "capitalist rationality" of the giant banks and their political cronies who led the world economy off a cliff, a role that has not won him or his bank many friends in the government. As Andrew Redleaf and Richard Vigilante pungently comment in *Panic*:

Left to their own devices, banks tend to be crabby and cautious with their money; those that aren't tend to die quick and painful deaths. Governments don't like crabby, cautious, or for that matter, independent banks. Governments like banks that are free and easy with their money and responsive to the governments' desires. Governments prefer loose credit to tight credit, and best of all they like loose credit extended to government's friends and pet projects, such as expanding home ownership....[2]

The government's favorite banks, therefore, were the Federal National Mortgage Association ("Fannie Mae") and the Federal Home Loan Mortgage Corporation ("Freddie Mac"), which were created to promote home ownership. Posner describes these monsters as private institutions, but they are "government-sponsored enterprises" (GSEs)—chartered and championed by the government—and they shape and stultify the entire mortgage system. Through their targeted purchases in the secondary market, they stand between mortgages and their actual issuers and owners, obscuring the information on which those mortgages are based. Because Fannie and Freddie are ostensibly private, their liabilities do not appear on the federal balance sheet, but they enjoy the implicit backing of the U.S. Treasury. At the end of 2009 these bureaucracies commanded $8.1 trillion of guaranteed debt, a staggering sum when compared with the $7.8 trillion of total marketable debt for the entire U.S. government.

The giant Wall Street banks, the federal piper's obedient dancers, were not far behind the GSEs in government favor. Wilmers acerbically questions the notion that any bank is "too big to fail. Looking at their performance, and the vast bailouts they required, perhaps they are too big to succeed." Their only strategy was the legally dubious conversion of government guarantees and insurance schemes into egregious levels of leverage. These tactics could transform trivial, sub-penny gains in rapid-fire stock trading into money-printing orgies. In early 2012, Wilmers pointed out that nine billion dollars of Goldman Sachs's heralded profits came not from profitable investments but from the appreciation of its debt as a result of the government's zero-interest-rate policy.

Wilmers contends that the Dodd-Frank Act, with 2,323 pages and 217 new regulations, enshrines the government's implicit backing of the established Wall Street banks despite their having strayed far beyond legitimate banking services. The dance goes on.

"It has become a virtual casino," Wilmers told the *New York Times'* Joe Nocera. "To me, banks exist for people to keep their liquid income, and also to finance trade and commerce." Yet 75 percent of the 2010 revenues of the five biggest Wall Street banks came from highly leveraged trading and money shuffling rather than from lending or even from conscientious investment.[3] In fact, since trading revenues go straight to the bottom line, while these banks have all too often been losing money on loans, all their income essentially comes from trading. Providing government-guaranteed funds and mortgages as safety nets for these financial funambulists has no theoretical justification. But dancing Jamie Dimon dismisses bankers like Wilmers as reactionaries who would return to the horse and buggy, which is how Dimon characterizes making capitalist loans to industries you understand.

Meanwhile, nothing at all has been done to rein in the chief miscreants in the mortgage scandal, Fannie Mae and Freddie Mac. The federal government had to pump in $126 billion when these otiose operations were forced into conservatorship—the GSE version of bankruptcy. The Congressional Budget Office estimates that including the operating costs of these GSEs in the federal deficit would increase it by $291 billion and predicts another $200 billion in new burdens in coming years.

Wilmers is particularly galled by the damage that Fannie and Freddie inflicted on private investors. Treasury Secretary Henry Paulson forced Main Street banks, including M&T, to acquire preferred shares in the GSEs just before the mortgage crisis erupted. That stock, whose value depended on the issuance of regular dividends, dropped to two cents a share. The compliant banks absorbed a loss from Fannie and Freddie of $1.8 billion, and several regional banks went broke. The giant Wall Street banks, meanwhile, garnered $169 billion in help from the Troubled Asset Relief Program (TARP). Acknowledging that the largest institutions have since paid back most of their TARP funds, Wilmers points out that "this assistance allowed them to emerge from the crisis in a stronger competitive position," in contrast to their smaller counterparts that actually make their money from banking.

Pouring all of this money into the real estate market and its rancid political dependencies corrupts the entire system. Residential housing is mostly consumption rather than investment. Yet Wilmers points out a devastating fact: without any version of Fannie and Freddie and without even a tax deduction for mortgage interest, Canada has a higher rate of home ownership than the United States does. The National Bureau of Economic Research finds that Fannie and Freddie actually *reduce* home ownership by their distortion of the markets and the resulting increase in the volatility of housing prices. In other words, Fannie and Freddie are utterly useless.

Six hundred miles south of Buffalo, in the small city of Wilson, North Carolina, is a bank about twice as big as M&T that also stayed away from the subprime punchbowl. Founded in 1872 by two young Confederate veterans, Alpheus Branch and Thomas Jefferson Hadley, trying to do something entrepreneurial in the wreckage of the Civil War, Branch Banking and Trust was the only bank of eight in its hometown to survive the stock market crash of 1929.

In 1971, near the bank's centennial and in the twilight of the Vietnam War, John Allison, a recent University of North Carolina graduate with a countercultural fondness for the ethics of Aristotle and Ayn Rand, went to work at BB&T in Wilson. It was still a farm bank, run by an aging group of managers who "were finally hiring some young people," as Allison recalled to Stephen Hicks in an interview forty years later in *Kaizen*. He had an idea of becoming a financial lawyer and thought that in such a small bank he could "make more of a difference" before he went off to law school. The grandson of a fatherless newspaper boy who worked his way up in the same newspaper over fifty years and the son of a World War II veteran who had spent over forty years at the same telephone company, Allison continued the family tradition of stability, spending the next thirty-seven years at BB&T. ("I really liked banking," he told Hicks, "and I did enough with lawyers to know I wouldn't enjoy being an attorney.")

"Mr. Allison of BB&T has the tall, lean frame, copper-colored hair and confident demeanor of many of Ms. Rand's fictional heroes," reported the *New York Times* in 2009. Others noticed the likeness as well. Donald Luskin and Andrew Greta featured Allison in their 2011 book, *I Am John Galt,* which profiles real-life versions of Randian heroes and villans—figures like Steve Jobs (a hero) and Paul Krugman (a villain).[4] *Forbes* drew the same conclusion, except that Allison "lacks the swagger." This matter-of-fact leadership would have a powerful effect on the bank where he worked.

Soon after Allison arrived at BB&T, Duke offered him a scholarship to its business school. Allison committed himself to the three-hour round trip between Wilson and Durham as a student and part-time banker—teller, bookkeeper, learner—season after season driving back and forth around Raleigh. When money grew tight, he picked up more hours at the bank and kept driving to Duke in the evenings. He and his immediate boss discovered they shared an enthusiasm for the work of Ayn Rand, and they turned the aimless "management development" program into, among other things, an *Atlas Shrugged* reading group.

This young group—inspired by the idea of making the world a better place by being the best bankers they could be—launched "an unintended revolution" in 1980, Allison told Hicks, beginning to influence how the bank was run. By 1987, the thirty-nine-year-old Allison was president, closely supported by four of his fellow *Atlas Shrugged* readers, and two years later he was CEO and chairman.

The bank, which had more than tripled in size throughout the Carolinas in the preceding two decades, was "financially sound," Allison remembered, "but we were almost certainly destined to be acquired in the consolidating industry unless we changed our strategy." As the savings and loan crisis rolled on, the new president and his team recognized that "there were a number of healthy thrifts that had stayed in the traditional bread-and-butter home loan business and had a lot of discipline. They were getting killed by the negative publicity going on about the industry." BB&T, confident that the law prohibiting banks from buying thrifts was about to change out of necessity, kept doing its research, and "we actually announced the first thrift acquisition by a commercial bank literally before the law changed."

This approach of coupling the big picture to research on the ground kept BB&T on the right path again and again, as it kept making careful acquisitions throughout the end of the twentieth century. The company culture of Aristotelian ethics and Randian purpose—carefully cultivated, and transplanted to each acquisition—paid off again and again. BB&T's response to the "pick-a-payment" negative amortization mortgages that appeared early in the twenty-first century is characteristic. The head of the mortgage department, Tim Dale, refused to make them on moral grounds. Allison explained that they would say to the demanding customers and to the regulators: "Look, this is how we run our business, this is a long-term perspective. We believe that some of these things that are going on are going to be detrimental. I can't prove that to you. It might be ten years before it happens, but we run our business based on principles."

Like M&T, BB&T made it through the sub-prime crisis without a single quarterly loss. "We avoided all the major excesses and irrationalities of the industry," Allison writes in his 2013 book, *The Financial Crisis and the Free Market Cure.* (He notes wryly that "most of the CEOs of major banks in 2007 had not been CEO in 1990 and did not know how bad business can get.")[5] When he heard rumblings about TARP, he went to Washington to try to stop it. "Paulson and [Fed chairman Ben] Bernanke were in almost constant conversation with the unhealthy institutions, but they would not talk with the leaders of the healthy banks," he remembered.

Bernanke and Paulson had their own motivations for TARP, as Allison explains in his book. A student of the Great Depression, the chairman of the Federal Reserve was trying to avoid the disasters of the 1930s, when the government's efforts to save individual banks actually destroyed them by provoking runs. Bernanke concluded in 2008 that he must cast his saving tarpaulin over the healthy and desperate alike. On the other hand, the secretary of the Treasury, a former investment banker, "realized that by 'encouraging' the healthy banks to participate in TARP, he would make the program appear to be more successful for taxpayers."

The closed doors in Washington were the last straw for Allison in his career-long struggle with misguided regulation. Ready to pass the torch to the younger generation of bankers he had spent his career cultivating, Allison had already decided to retire at the end of 2008, but he hadn't expected to go out this way. In October, TARP passed. As Allison describes in his book, BB&T got a velvet-gloved tap on the shoulder from their regulators the next day. "The essence of the message was that although BB&T had substantially more capital than it needed under long-established regulatory standards, given the current economic environment, the regulators were going to create a new set of capital standards. They did not know what the standards would be. However, they were 'very concerned' that we would not have enough capital under these new standards unless we took the TARP capital. They had a regulatory team in place to reexamine our capital position immediately unless we took the TARP funding.

"The threat was very clear. We said, 'Please sign us up for TARP.'" And thus began Allison's final month as CEO.

"Several months later... BB&T passed the 'stress test' with flying colors." They had been forced to pay an above-market interest rate for money that they did not need. As soon as repayment was allowed, BB&T was the first to return the money, "(admittedly by only a few minutes)," Allison writes. The financial data firm Morningstar chose Allison as runner-up to Warren Buffett in their analysis of CEO of the year.

In Allison's nineteen years as CEO, BB&T grew from a statewide bank worth $4.5 billion to a $152 billion bank operating in eleven southern states and the District of Columbia. When Stephen Hicks asked if he didn't find this extraordinary, Allison answered in terms of knowledge and power. "We had invested very heavily over a long period of time in [the education of] quality leaders.... For example, we had really strong community bank presidents. They had a much higher level of authority than our competitors" and "were held very responsible—they owned the process." But this decentralized power, he explained, had to be guarded. "Centralized systems are a natural process because elitists consolidate power (they don't think of themselves as elite—these are well-intended people who like to run their system) and they hate deviations." So as CEO, Allison "had to constantly go back to the home office... and say:

'Yes, I understand that this makes your job harder but here are the benefits of having this localized decision-making: We get better information, we get faster decisions, we understand the market better.'"

BB&T remained successful during the crisis, says Allison, because "we had much of our authority in our community banks. We had better information, and that's non-trivial." He repeats in interview after interview: "The most important human resource, the only true natural resource, is the human mind."

The idea that Wall Street's dancing princes possess some exotic expertise to which the rest of the market must attend is ludicrous. Their only exceptional capability is lobbying the government and lavishing bonuses on themselves. Many have moved on from lobbying the government to running the Treasury, where they have done no better than they did running the banks.

Meanwhile, Wilmers remains in charge of M&T and has become rich as a result of the appreciation of his shares, now worth some $80 million. In the current debate over the future of capitalism, there is widespread sentiment to take that wealth away from him. Allison has become a "Distinguished Professor of Practice" at Wake Forest University's business school and the president of the Cato Institute, a Randian redoubt in Washington. From his perspective, this is just an extension of his time at BB&T: "When you see a light go on, that's a very satisfying thing. And I saw a lot of lights go on in my career."

Reviewing Allison's book in the *Weekly Standard*, Lewis Lehrman summed up his assessment of the banking crisis:[6]

> Here we see a financial malignancy metastasize in the hands of hapless surgeons in Congress, in the White House, and in the Treasury and the Federal Reserve.

> The only way there could have been a bubble in the residential real estate market was if the Federal Reserve created too much money. It would have been mathematically impossible for

a misinvestment of this scale to have happened without the monetary policies of the Fed.

· · · ·

In the lead-up to the collapse, Greenspan "created a structure of negative real interest rates," forcing rates down to 1 percent, encouraging and providing incentives to banks and borrowers to buy and sell poisoned products—and to take on vast amounts of shaky debt and leverage which could not be sustained if higher real interest rates returned.

Enter Ben Bernanke, former Princeton economist and Fed vice chairman under Greenspan: "After he became chairman [in 2006] Bernanke rapidly raised interest rates and created an inverted yield curve" (higher short-term rates than the high long-term rates). But homeowners, businesses, and banks, lured by the Fed into leverage and cheap loans, could not finance the Greenspan interest-rate increases, followed by Bernanke's abrupt move to raise the federal funds rate (the interest rate the Fed charges banks) to 5.25 percent.

The impact of such an interest rate move (from 1 percent to 5.25 percent) must be thought of as a price increase of 500 percent plus—similar to an increase in the price of a loaf of bread from $2.50 to $15. Remember that Americans (and consumers worldwide) had borrowed and leveraged themselves during the period in which Alan Greenspan forced the federal funds rate down to 1 percent....

Then Bernanke "held the inverted yield curve for more than a year (from July 2006 to January 2008), one of the longest yield-curve inversions ever." And inverted yield curves historically lead to recessions. In a word, according to Allison, the Federal Reserve was both the fundamental cause of the real estate bubble and the agent of its collapse. But Bernanke "was adamant that there would not be a recession."

Allison's conclusion?

> In my career, the Fed has a 100 percent error rate
> in predicting and reacting to important economic
> turns... [because it] is trying to arbitrarily set the
> single most important price in the economy—the
> price of money.

> And yet, as we know, setting wage and price controls, from
> the time of Diocletian to Richard Nixon, has proven in
> every case a disaster for economies and the people
> entrapped by them.[7]

In pithy and persuasive prose, Allison and Wilmers—two good bankers—provided the most powerful explanations of the crisis. Yet it is the dancing princes and crony capitalists who remain in control of American banking and monetary policy.

The One Percent

As far as millionaires and billionaires are concerned,
they're experiencing a horrifying revolution:
consumption equality. For the most part, the wealthy
bust their tail, work 60–80 hour weeks building some
game-changing product for the mass market, but
at the end of the day they can't enjoy much that
the middle class doesn't also enjoy. Where's the fairness?
What does Google founder Larry Page
have that you don't have?
Luxury suite at the Super Bowl? Why bother?
You can recline at home in your massaging lounger
and flip on the ultra-thin, high-def,
55-inch LCD TV you got for $700....
The greedy tycoon played by Michael Douglas had
a two-pound, $3,995 Motorola phone in the origi-
nal "Wall Street" movie.... Given its memory, today's

32-gigabyte smartphone would have cost $1 million
back then.... But no one could build it—... it wasn't
until there was a market for millions of smartphones
that there was a market at all....
Medical care? Thanks to the market, you can afford
a hip replacement and extra-capsular cataract extraction
and a defibrillator—the costs have all come down
with volume. Arthroscopic, endoscopic,
laparoscopic [surgery], drug-eluting stents...wouldn't
have been invented to service only the 1%.
I admit that a private jet beats the TSA rub-a-dub....
But thanks to guys like Richard Branson and
airline overbuild, you can fly almost anywhere in
the world for under $1,000....
Spot the pattern here? Just about every product or
service that makes our lives better requires a mass market
or it's not economic to bother offering. Those who invent
and produce for the mass market get rich.... Income
equality may widen, but consumption equality
will become more the norm....
Compared to 20 years ago, or even five years ago,
chances are that you're richer. Try to enjoy it.

—ANDY KESSLER, "THE RISE OF CONSUMPTION EQUALITY,"
WALL STREET JOURNAL, JANUARY 3, 2012

IN THE FIRST years of the second decade of the new Millennium, the carnival of chaos and self-serving on Wall Street and in Washington provoked a new debate over the future of capitalism. As always, whenever the government botches the economy, whenever recession strikes and growth languishes, the media and the academy strike out against newly discovered and invariably misrepresented inequality.

The idea is that the difference between people is what they own rather than what they know. But wealth is valuable only if combined with information. If it wears the blinders of ignorance or is corroded by the acid of greed it quickly dissolves, like the fortunes of lottery winners or Vegas jackpot millionaires.

America's entrepreneurs live in a world with four billion poor people. Mostly unarmed men and women, these leaders of business command little political power or means of defense. Democratic masses or military juntas could take their wealth at will. Why, on a planet riven with famine, poverty, and disease, we might ask, should this tiny minority be allowed to control riches thousands of times greater than their needs for subsistence and comfort? Why should a few thousand families command wealth far exceeding the endowments of most nations?

More specifically, why should Mark Zuckerberg, the proprietor of Facebook, command a fortune estimated at $18 billion while Suzie Saintly, the social worker, makes a mere forty thousand a year? Or why should Bill Gates, Microsoft founder, be worth over $50 billion while Dan Bricklin, the inventor of the pioneering VisiCalc spreadsheet that launched the personal computer into business applications, be reduced to tilling his tiny consultancy "Software Garden" with a handful of colleagues? Why should Warren Buffett be permitted to parlay his positions in newspapers, insurance companies, and Coca-Cola into a fortune that sometimes makes him the world's richest man? Why should coffee monger Howard Schultz enjoy a $5 billion fortune while Harry Homeless lives on a rug on a grate? And why should Michael Milken, the junk bond pioneer who ended up in prison for trumped up "economic crimes," still be a billionaire, while the president of the United States earns a puny $400,000 per year?

Does any of this make any sense?

In statistical terms, the issue arises just as starkly. Why should the top 1 percent of American households have a net worth higher than the bottom 90 percent, and the top 1 percent of income earners take in more before taxes than the bottom 50 percent? The bottom 30 percent, awash in debt and dependent on government, have no measurable net worth at all.

On a global level, the disparity assumes a deadly edge. Why should even this bottom fifth of Americans be able to throw away enough food to feed a continent while a million Rwandans die of famine? Why should the dogs and cats of America eat better than the average citizen of this unfair planet?

Jeffrey Sachs, the director of the Earth Institute at Columbia University, contends that "the three main aims of an economy [are] efficiency, fairness, and sustainability," and maintains that the U.S. economy is failing on all three. He sees the fundamental problem as fairness. "Too many of America's elites—among the super-rich, the CEOs, and many of my colleagues in academia—have abandoned a commitment to social responsibility. They chase wealth and power, the rest of society be damned.... [T]he top one percent of American households...sit at the top of the heap at the same time that around 100 million Americans live in poverty or in its shadow.... Our greatest national illusion," he writes, "is that a healthy society can be organized around the single-minded pursuit of wealth."[1]

We all know that life is not fair, but many people believe that these huge disparities defy every standard of proportion and propriety. They apparently correspond neither to need, nor to virtue, nor to IQ, nor to credentials, nor to education, nor to social contribution.

Most observers now concede that capitalism generates prosperity, at least most of the time. But the rich, who are its poster children, hardly make it attractive. Look at the "*Forbes* 400" list of America's wealthiest people, for example, and hold your nose (no movie stars here). Many of them are short and crabby, beaked and mottled, fat and foolish. Several never finished high school, and only about two-thirds of those who went to college managed to graduate. A society may tolerate an aristocracy certified by merit. But capitalism seems to exalt strange riffraff with no apparent rhyme or reason.

Couldn't we create a capitalist system without fat cats and predatory bond traders? Wouldn't it be possible to contrive an economy that was just as prosperous, but with a far more just distribution of wealth? In *The European Dream: How Europe's Vision of the Future Is Quietly Eclipsing the American Dream*, Jeremy Rifkin writes that Europeans have already achieved this goal, enjoying "a longer life span and greater literacy, and...less poverty and crime, less blight and sprawl, longer vacations and shorter commutes to work.... While the American Dream emphasizes unrestrained economic growth, personal wealth and the pursuit of individual self interest," Rifkin explains, "the European Dream focuses more on sustainable development, quality of life, and the nurturing of community.... Europeans place more of a premium on leisure and even idleness."[2] Writing in 2004, Rifkin thought that Europe's success was attributable to its greater commitment to equality.

In a celebrated tract, *The Spirit Level: Why Equality is Better for Everyone*, the British social scientists and epidemiologists Richard Wilkinson and Kate Pickett present evidence for the assumption that fairness is essential to efficiency and happiness.[3] They find that almost every aspect of human welfare—from life expectancy to mental illness, from violence to illiteracy, from infant mortality to homicides, from obesity to imprisonment, from teenage births to "level of trust"—is determined not by the wealth of the society but its equality. With a formidable array of charts, they assert that societies like the United States with a larger gap between the rich and the poor make all citizens unhappy, sick, obese, addictive, and feckless, including the rich.[4]

Using a demand-side model that focuses on consumption rather than production, happiness rather than innovation, *The Spirit Level* sets up countries such as Sweden and Japan, happy and prosperous, as foils for the United States, Israel, and Singapore, which lead the world in creation of new wealth and technology. Wilkinson and Pickett neglect to mention that without the new wealth and technology, the entire globe would sink into famine and plague.

The statistical analysis in *The Spirit Level* is shoddy, mixing correlations and causations without scruple or controls. Most of the happy nations are small, and small nations tend to have smaller variances of all kinds. Japan is large but homogeneous and benefits from American defense spending. America is large, varied, and unequal, and the source of food and technology for the world. Swedes and Japanese, among other favored groups in the book, enjoy the same health and longevity while living in the United States as they do in their home countries. Israel is small and unequal and the global leader in innovation and invention. According to *The Spirit Level,* the United States and Israel are two of the world's most miserable countries, lagging behind Cuba and Greece. The book presumes that leisured "happiness" is the goal of human life. The Greeks may have enjoyed their two-decade vacation while they fed on the dynamism of the world economy. But without the dynamism of the world economy the Greeks would not even survive.

This case against capitalism is based on a critique of the character of capitalists. It depicts the system as a Faustian pact, by which we trade greed for wealth. The economy becomes a materialist mechanism, deterministically inducing productive behavior through "incentives," readily modeled for a computer simulation. Human beings become machines, too, ruled by the pursuit of pleasure, which is manifested in the quest for material goods and sensual satisfactions. The whole picture ultimately derives from the long-discredited "Skinner box" concept of human psychology as governed by stimulus and response.

According to this view, which pervades American universities, the world would be a better place if rich entrepreneurs saw their winnings capped at, say, $20 million. Surely Sam Walton's heirs could make due on a million dollars or so a year of annual income, four or five times the salary of the president. Sam himself, after all, is long gone. Why should his descendants thrive on his gains? How is a system that transfers wealth through the happenstance of inheritance either moral or meritocratic?

Most defenders of capitalism reply that the inequalities we complain of are an unavoidable reflection of the processes that create wealth. These processes thrive when the private sector controls most of the wealth and economic decision-making. In any society without large inheritances, wealth and power would eventually gravitate to government control. Arbitrary and coercive bureaucrats might impose a sterile egalitarian utopia ruled by politicians and their constabulary. Economic and technological progress would stagnate, and the economy would become a zero-sum game dominated by politicians. This is a plausible argument against forced equality, implying that capitalism doesn't make sense, morally or rationally, but it does make wealth, and wealth is essential for the very survival of a planet with an ever-growing population. So, they say, don't knock it.

This standard case for capitalism maintains that greed may drive Angelo Mozilo of Countrywide Financial or the Ponzi schemer Bernard Madoff to criminal behavior. But, the argument runs, greed also makes the system go. Because greed is less trammeled in the United States than in Ethiopia, Harry the Homeless on the grate eats better than the middle class of Addis Ababa.

This was essentially the argument of Adam Smith, the first and still most-quoted apologist of capitalism. He declared that it is only from the entrepreneur's "luxury and caprice," his desire for "all the different baubles and trinkets in the economy of greatness," that the poor "derive that share of the necessaries of life, which they would in vain have expected from his humanity or his justice." In perhaps his most famous lines, Smith wrote of entrepreneurs:

> In spite of their natural selfishness and rapacity, though they mean only their own conveniency, though the sole end which they proposed, from the labours of all the thousands they employ, be the gratification of their own vain and insatiable desires...they are led by an invisible hand...and without intending it, without knowing it, advance the interest of society.

Thus capitalism's greatest defender wrote of the rich of his day. More recent economists, from Paul Krugman to Robert Reich to Charles Ferguson, speak of the rich wallowing in their riches and implicitly bilking the poor of the necessities of life. They want to "save capitalism," yet again, by raising tax rates on the rich and forcibly redistributing incomes.

Yet these arguments fail to come to terms with the actual behavior of entrepreneurs. Despite the opportunity for indolence, most of them work fanatically hard. In proportion to their holdings, their output, and their contributions to the human race, they consume less than any other group of people in the history of the world.

Far from being greedy, America's leading entrepreneurs—with some exceptions—display discipline and self-control, hard work and austerity, excelling those found in any college of social work, Washington think tank, or congregation of bishops. They are a strange collection of riffraff to be sure, because they are chosen not according to blood, credentials, education, or services rendered to the establishment. They are chosen for performance alone, for service to the people as consumers.

Greed is an appetite for unneeded and unearned wealth and power. Because the best and safest way to gain unearned pay is to persuade the state to take it from others, greed leads, as by an invisible hand, toward ever more government action—toward socialism, government subsidies, and guarantees, all utterly inimical to the morality of capitalism.

Socialism in all its forms—from Wall Street subsidy-seekers to bureaucratic profiteers—is in practice a conspiracy of the greedy to exploit the productive. The beneficiaries of the government's transfers of wealth and income smear their betters with the claim of avarice that they themselves deserve. The rich in general have earned their money by contributions to the commonweal that far exceed their incomes, or have inherited their fortunes from forebears who did likewise. What is more, most entrepreneurs continue their work to enrich the world.

Greed is actually less a characteristic of Larry Page, Mark Zuckerberg, or Warren Buffett than of Harry the Homeless. Harry may seem pitiable. But he and his advocates insist that he occupy—and devalue—some of the planet's most valuable real estate. From the beaches of Santa Monica to San Francisco's Presidio to Manhattan's Central Park, he wants to live better than most of the population of the world throughout human history, but he does not want to give back anything whatsoever to the society that sustains him. He wants utterly unearned wealth. That is the essence of avarice. If you want to see greed in action, watch Jesse Jackson regale an audience of welfare mothers on the "economic violence" of capitalism, or listen to leftist college professors denouncing the economic system that provides their freedom, tenure, long vacations, and other expensive privileges.

Most of America's entrepreneurs bear no resemblance to the plutocrats of socialist and feudal realms, who get government to steal their winnings for them and then revel in their palaces with eunuchs and harems. The American rich, in general, cannot revel in their wealth because most of it is not liquid. It has been given to others in the form of investments. It is embodied in a vast web of enterprise that retains its worth only through constant work and sacrifice. Larry Page and Jeff Bezos live modestly. They give prodigally of themselves and their work. They reinvest their profits in productive enterprise that employs and enriches the world.

Nevertheless, the reason for the disparities between the 400 and the Four Billion is not that the entrepreneurs work harder or better or forgo more consumption. In dismissing the charge that the "1 percent of the 1 percent" indulge in a carnival of greed, we do not explain the real reasons for their huge wealth.

Since Adam Smith, a host of theories have been offered to answer the great enigma of capitalist inequality. There is the argument of rights: the creators of great wealth have a right to it. But the assertion of rights to vast fortunes created by thousands of people and protected by the state only restates the enigma in more abstract terms. Then there is the argument of carrots and sticks. Mark Zuckerberg's billions gave him a necessary incentive in the imperial expansion of Facebook and inspire others to work hard. But the critics can plausibly answer, "Sure, we all need incentives...but $14 billion at age 30?"

All of these arguments are beside the point. Capitalism's distribution of wealth makes sense, but not because of the virtue or greed of entrepreneurs, or the invisible hand of the market. The reason is not incentives, or carrots and sticks, or just deserts. The reason capitalism works is that the creators of wealth are granted the right and burden of reinvesting it—or choosing the others who receive it in the investment process.

Warren Buffett worried that his average personal tax rate of 17 percent was "unfair" when his secretary pays more. The accuracy of his 17-percent calculation aside (it ignores the 39-percent corporate income tax, the 55-percent inheritance tax, and other levies), the reason Buffett pays 17 percent has nothing to do with "fairness." He pays low rates because of the superiority of his entrepreneurial knowledge as an investor of capital. Buffett does not consume his income; he faces the challenge and onus of reinvesting his profits. Capitalism prevails because it assigns this exacting task to people like Warren Buffett rather than to people like his secretary—let alone to people in government who are driven by political rather than economic considerations.

Like a plant in a garden, an economy grows by photosynthesis—without the light of new knowledge and the roots of ownership, it withers. In general, wealth can grow only if the people who create it control it. Divorce the financial profits from the learning process and the economy stagnates.

The riches of the 400 all ultimately stem from this entrepreneurial process. Seventy percent of them received no inherited wealth to speak of, and most of the rest gained their fortunes from entrepreneurial parents.

Entrepreneurial knowledge has little to do with the certified expertise of an advanced degree from an establishment school. It has little to do with the gregarious charm of the high school student voted most likely to succeed. The fashionably educated and cultivated spurn the kind of fanatically focused learning undertaken by the 1 percent. Wealth all too often comes from doing what other people consider insufferably boring or unendurably hard.

The treacherous intricacies of building codes or garbage routes or software languages or groceries, the mechanics of butchering sheep and pigs or frying and freezing potatoes, the mazes of high-yield bonds and low-collateral companies, the murky arcana of petroleum leases or housing deeds or Far Eastern electronics supplies, the ways and means of pushing pizzas or insurance policies or hawking hosiery or pet supplies, the multiple scientific disciplines involved in fracking for natural gas or tapping shale oil or contriving the ultimate search engine, the grind of grubbing for pennies in fast-food unit sales, the chemistry of soap or candy or the silicon-silicon dioxide interface, the endless round of motivating workers and blandishing union bosses and federal inspectors and the IRS and EPA and SEC and FDA—all are considered tedious and trivial by the established powers.

Most people think they are above learning the gritty and relentless details of life that allow the creation of great wealth. They leave it to the experts. But in general, you join the 1 percent of the 1 percent not by leaving it to the experts but by creating new expertise, not by knowing what the experts know but by learning what they think is beneath them.

Entrepreneurship is the launching of surprises. What bothers many critics of capitalism is that a group like the 1 percent is too full of surprises. Sam Walton opens a haberdashery and it goes broke. He opens another and it works. He launches a shopping center empire in the rural South and becomes for a while America's richest man selling largely Chinese-made goods to Americans. Howard Schultz of Starbucks makes a fortune out of coffee shops, leaves, and watches his company decline in his absence. He returns and restores it to supremacy as a multifarious supplier of drinks, food, and home comforts outside of home. Herb Kelleher leaves the Northeast to become a lawyer in Texas. On the proverbial napkin he outlines plans for a new kind of airline in Texas. Defying the deepest beliefs of the experts in the established airlines—their gouge-and-gotcha pricing, hub-and-spoke routing, and diversity of aircraft sourcing—Kelleher builds Southwest Airlines. Bringing bus-like convenience, singing stewardesses, and business innovations, he creates the world's leading airline and a fortune for himself. Rather than retiring, he becomes chairman of the board of directors of the Federal Reserve Bank of Dallas.

This process of wealth creation is offensive to levelers and planners because it yields mountains of new wealth in ways that could not possibly be planned. But unpredictability is fundamental to free human enterprise. It defies every econometric model and socialist scheme. It makes no sense to most professors, who attain their positions by the systematic acquisition of credentials pleasing to the establishment above them. By definition, innovations cannot be planned. Leading entrepreneurs—from Sam Walton to Mike Milken to Larry Page to Bill Gates—

did not ascend a hierarchy; they created a new one. They did not climb to the top of anything. They were pushed to the top by their own success. They did not capture the pinnacle; they became it.

This process creates wealth. But to maintain and increase it is nearly as difficult. A pot of honey attracts flies as well as bears. Bureaucrats, politicians, bishops, raiders, robbers, revolutionaries, short-sellers, managers, business writers, and missionaries all think they could invest money better than its owners. Aspiring spenders—debauchers of wealth and purveyors of poverty—besiege owners on all sides in the name of charity, idealism, envy, or social change. In fact, of all the people on the face of the globe, it is the legal owners of businesses who have the clearest interest in building wealth for others rather than spending it on themselves, and it is usually only the legal owners who know enough about the sources of their wealth to maintain it.

Leading entrepreneurs consume only a small portion of their holdings. Nearly always they are owners and investors. As owners, they are the ones damaged most by mismanagement or exploitation or waste of their assets.

As long as Larry Page and Sergey Brin are in charge of Google—and recall the sun-drenched wisdom of Eric Schmidt—it will probably grow in value. But if you put Harry Homeless in charge of Google—or if, as Harry's compassionate proxy, you put a government bureaucrat in charge—within minutes the company would be worth a giggly half its former value. As other firms, such as Netflix and Starbucks, discovered from their missteps, an ascendant company can flush away much of its worth in minutes if fashions shift or investors distrust the management.

As a Harvard Business School study by Michael Jensen showed, even if you put "professional management" at the helm of great wealth, value will grow less rapidly than if you give owners the real control. A manager of Google might benefit from stealing from it or from turning it into his own cushy preserve, making self-indulgent "investments" in company planes and playgrounds and favored foundations that are in fact his own disguised consumption. It is only Page and Brin who would see their own wealth dissipate if they began to focus less on their customers than on their own consumption. (They may make this mistake anyway.) The key to their wealth is their resolution neither to spend nor to abandon it. In a sense they are as much slaves as masters of the Googleplex.

Even if it wished to, the government could not capture America's wealth from its 1 percent of the 1 percent. As Marxist despots and tribal socialists from Cuba to Greece have discovered to their disappointment, governments can expropriate wealth, but they cannot appropriate it or redistribute it. Similarly in the United States, a leftward administration can destroy the value of the 1 percent's property, but cannot seize it or pass it on.

Under capitalism, wealth is less a stock of goods than a flow of ideas and entropy. Joseph Schumpeter propounded the basic rule when he declared capitalism "a form of change" that "never can be stationary." The landscape of capitalism may seem solid and settled and thus seizeable; but capitalism is really a "noosphere"—a mindscape—as empty in proportion to the nuggets at its nucleus as the solar system is in proportion to the size of the sun. Volatile and shifting ideas, not heavy and entrenched establishments, constitute the source of wealth. No bureaucratic net can catch the fleeting thoughts of Eric Schmidt of Google, Jules Urbach of OTOY or Christopher Cooper of Seldon Technologies.[5]

Nonetheless, in this mindscape of capitalism, all riches finally fall into the gap between thoughts and things. Governed by mind but embodied in matter, to retain its value an asset must produce an income stream that can be expected to continue. The expectation can shift as swiftly as thought, but the things, alas, are all too solid and slow to change.

Warren Buffett's conglomerations of Coke and commercial paper, Nathan Myrvold's patent trove, John Paulson's hedges, Mark Zuckerberg's Facebook mansions, Intel's wafer fabs, the Crow family's real estate empire, the gilded towers of Donald Trump—all could become shattered monuments of Ozymandias tomorrow. "Look on my works, ye Mighty, and despair!" wrote Shelley in the voice of the king whose empires became mere mounds in the desert sands.

Like the steel mills of Pittsburgh, the railroad grid of New England, and the thousand-mile Erie Canal; like the commercial real estate of

Detroit and the giant nuclear plants and great printing presses of yester-year; like Kodak's photography-patent portfolio of a decade ago, the HP computers of a year ago, or the sartorial rage of last week; the material base of the 1 percent of the 1 percent can be a trap, not an enduring fount of wealth. In all these cases the things stayed pretty much the same. But thoughts about them changed. Much of what was supremely valuable in 1980 was nearly worthless by 2013.

Overseas interests could buy the buildings and the rapidly obsolescing equipment and patents of high-technology firms. But the leadership, savvy, and loyalty would probably be lost in the sale. If the Chinese or Arabs bought all of Silicon Valley, for example, they might well do best by returning it to the production of apricots, oranges, and bedrooms for San Francisco. Capturing the worth of a company is incomparably more complex and arduous a task than purchasing it.

In the Schumpeterian mindscape of capitalism, the entrepreneurial owners are less captors than captives of their wealth. If they try to take it or exploit it, it tends to evaporate. As Bill Gates, already a deci-billion-aire on paper, once put it during his entrepreneurial heyday, "I am tied to the mast of Microsoft." If Gates had tried to leave or substantially cash out at any time during the early decades, the company would have plum-meted in value more rapidly than he could have harvested the funds. When the founders of Bain & Company attempted to cash out in 1991, taking $200 million with them, Mitt Romney was faced with the likely bankruptcy of the firm. He had to confront a Goldman Sachs effort to close it down. As the once-lucrative firm collapsed, Romney cut the share of his departing partners in half in order to save his company and his reputation in business. David Rockefeller devoted a lifetime of sixty-hour weeks to his own enterprises. Younger members of the family wanted to get at the wealth and forced the sale of Rockefeller Center to Mitsubishi. But they will discover that they can keep the wealth only to the extent that they serve it, and thereby serve others rather than themselves.

Most of America's leading entrepreneurs are bound to the masts of their enterprises. They are allowed to keep their wealth only as long as they invest it in others. It has been given to others in the form of investments. It is embodied in a vast web of enterprises that retains its worth only through constant work and sacrifice. In a real sense, they can keep only what they give away. Capitalism is a system that begins not with taking but with giving to others.

In feudal or socialist or communist backwaters, a register of material wealth could accurately depict its distribution. There, riches reside chiefly in land, natural resources, police powers, and party offices, often held in perpetuity. Under socialism the 1 percent might represent a dominant establishment, combining both political and economic clout.

The abiding delusion of economic managers is that they can guarantee the value of things rather than the ownership of them. Ownership bears consequences, whether profits or losses. The great mistake of the Bush and Obama administrations' response to the crisis of 2008 was to shield the owners from the costs of their mistakes. By guaranteeing things, government tends to destroy their value, which depends on dedicated ownership. In the United States, the Constitution guarantees only the right to property, not to its worth—or so it seemed until recently.

The belief that wealth subsists not in ideas, attitudes, moral codes, and mental disciplines but in identifiable and static things that can be seized and redistributed is the materialist superstition. It stultified the works of Marx and other prophets of violence and envy. It frustrates every socialist revolutionary who imagines that by seizing the so-called means of production he can capture the crucial capital of an economy. It is the undoing of nearly every conglomerateur who believes he can safely enter new industries by buying rather than by learning them. It confounds every bureaucrat who imagines he can buy the fruits of research and development. The cost of capturing technology is mastery of the knowledge embodied in the underlying science. The means of entrepreneurs' production are not land, labor, or capital but minds and hearts.

The reason for the huge wealth gap between Larry Page and Suzie Saintly, between Donald Trump and Harry Homeless, between Oprah and Obama, or between the 1 percent and any number of other worthy men and women, is entrepreneurial knowledge and commitment.

Keeping their wealth by investing it in others, they know how to maintain and expand their holdings, and the market knows of their knowledge. Thus they increase the wealth of America and the opportunities of even the poorest.

The wealth of America is not an inventory of goods; it is an organic, living entity, a fragile, pulsing tissue of ideas, expectations, loyalties, moral commitments, visions. To vivisect it for redistribution is to kill it. As President Mitterrand's French technocrats discovered in the 1980s, and as President Obama's quixotic American eco-crats are discovering today, the proud usurpers of complex systems of wealth soon learn that they are administering an industrial corpse, a socialized Solyndra.

Whatever the inequality of incomes, it is dwarfed by the inequality of contributions to human advancement. As the science fiction writer Robert Heinlein wrote, "Throughout history, poverty is the normal condition of man. Advances that permit this norm to be exceeded—here and there, now and then—are the work of an extremely small minority, frequently despised, often condemned, and almost always opposed by all right-thinking people. Whenever this tiny minority is kept from creating, or (as sometimes happens) is driven out of society, the people slip back into abject poverty. This is known as 'bad luck.'" President Obama unconsciously confirmed Heinlein's sardonic view of human nature in a campaign speech in Iowa: "We had reversed the recession, avoided depression, got the economy moving again, but over the last six months we've had a run of bad luck."

All progress comes from the creative minority. Even government-financed research and development, outside the results-oriented military, is mostly wasted. Only the contributions of mind, will, and morality are enduring. The most important question for the future of America is how we treat our entrepreneurs. If our government continues to smear, harass, overtax, and oppressively regulate them, we will be dismayed by how swiftly the engines of American prosperity deteriorate. We will be amazed at how quickly American wealth flees to other countries.

Most American entrepreneurs are staying in America. But the global telecommunications network allows them to invest their liquid funds overseas at the speed of light down a fiber-optic line. While young entrepreneurs once built the fortunes of future decades in America, they now begin them overseas—in China, India, and Israel—or fail to begin them at all. Their own worth and the wealth of the United States decline in the process. Other countries are beginning to thrive where America once flourished, as the secrets of American wealth spread across the increasingly global mindscape of capitalism.

Those most acutely threatened by the abuse of American entrepreneurs are the poor. If the rich are stultified by socialism and crony capitalism, the lower economic classes will suffer the most as the horizons of opportunity close. High tax rates and oppressive regulations do not keep anyone from *being* rich. They prevent poor people from becoming rich. High tax rates do not redistribute incomes or wealth; they redistribute taxpayers—out of productive investment into overseas tax havens and out of offices and factories into beach resorts and municipal bonds. But if the 1 percent and the 0.1 percent are respected and allowed to risk their wealth—and new rebels are allowed to rise up and challenge them—America will continue to be the land where the last regularly become the first by serving others.

PART THREE

The Future

The Black Swans
of Investment

PERHAPS THE MOST interesting and fashionable of financial philosophers and information theorists, Nassim Nicholas Taleb, is a libertarian, and in all likelihood he deems you a turkey. There he sits, in Paris or New York, in a "dilapidated but elegant café," as "unpolluted" as possible with "persons of commerce," watching you walk by. He sneers— but don't take offense: he views much of the passing world with disdain.

Maybe you are a red-tied banker, besotted with Republican beliefs. Taleb turns up his nose. Possibly you consider Vivaldi's *Four Seasons* a "refined" work of music. You have affronted one of his recherché aesthetic opinions. Possibly you find Michael Bloomberg one of the few rich people who actually conform to Taleb's generalization about the wealthy—"inelegant, dull, pompous, greedy, unintellectual, selfish, and boring."

Perhaps, on a deeper level, you try to plan ahead, forgo gratifications, and save for the future, pursuing the good life of home, church, enterprise, and family. Maybe this traditional code makes it easy for you to live with the possibility that ultimately the universe is futile and random. In most cases, Taleb sees you as human poultry.

Don't bridle, though, at Taleb's elegant scorn. He thinks turkeys have, for the most part, a good life. On the whole, he envies them. Pampered by attentive keepers, comfortable in their pens and sleek suits, preening over their red wattles, coddled by hens and chicks, these happy gobblers feel complacent about the world…right up to the day before Thanksgiving. Then, like the Federal Reserve serenely reigning over the "Great Moderation," they meet what Taleb calls their "black swan." That's the out-of-the-blue disaster that confounds all their seemingly reasonable expectations.

Taleb's key insight comes from the work of David Hume, and it's known as the inductive fallacy: no number of authoritative sightings of white swans can prove the absence of black swans. No succession of good days offers evidence against a bad day tomorrow—a Thanksgiving axe (if you're a turkey), a 9/11, a Pearl Harbor, a tornado, another financial meltdown, a nuclear explosion, a meteor.

More than a decade ago, Taleb published an autobiographical book called *Fooled by Randomness: The Hidden Role of Chance in Life and in the Markets*.[1] This stylish and intriguing account of his reflections as a "no nonsense" trader of derivatives became a surprise bestseller. That "unexpected success" (which I predicted, actually, when I read it) brought forth an ungainly child, one even more disdainfully Delphic and devoted to the same misleading ideas—*The Black Swan: The Impact of the Highly Improbable*.[2] Vaulting onto the *New York Times* bestseller list, it thrust the professor of the sciences of uncertainty onto the media stage, making his reputation as a belletristic cosmopolitan thinker just in time for the financial crisis of 2008.

Much of his book is a catalogue of the errors of logic afflicting all us turkey investors and pushing us toward the cliffs of unexpected catastrophe. He points especially to *survivorship bias*. Survivorship distorts any phenomenon in which only the survivors are around to be analyzed and included in the data. Taleb compares survivor bias with the effect of harsh conditions on their victims—whether gulag inmates said to have been "hardened" by the experience or irradiated rats that are declared to have become "super-rats" as a result. These punishments do not "harden" their victims but merely select the hardest as survivors.

Other temptations for the turkey include the narrative fallacy, which leads us to ascribe cause and coherence to a sequence of interesting or emotionally charged events. We tell stories and listen to them, as if they captured the deeper truths of life, when almost all the most savory anecdotes are too good to be true or too rare to be representative.

The narrative fallacy converges with the ludic fallacy: we turkeys are alleged to believe that the world is a game, played by the rules. The Thanksgiving axeman will not come because it is against the law of the land or against the law of probabilities. Even casino owners believe this fable. But Taleb demurs. He tells about a casino he visited and advised in Vegas, which turned out to have mastered all the possible threats from within the game: counting, cheating, or outwitting the house. But the actual misfortunes came from beyond the rules—an errant tiger that attacked its trainer (Roy, of Siegfried & Roy), a disgruntled builder who threatened to blow up the place, an IRS scam by an ingenious clerk. These black swans came near to bringing down the house.

Taleb believes that all ventures and markets resemble this casino and that human beings persistently underestimate the likelihood and significance of black swans. Perhaps he should leave his café more often. A perusal of the *New York Times* or any television news program—two sources of information that Taleb commendably shuns—reveals a propensity to panic over global warming, avian flu, the Chinese navy, asbestos, nuclear power plants, North Korea, radon, DDT, electromagnetic pulse, and other trumped-up perils. But Taleb wrings his hands over global warming and nuclear power anyway. Bestseller lists are filled with books on the approaching economic doom.

In fact, everybody knows about black swans, and as Ken Fisher shows in his investment classic, *The Only Three Questions That Count,*[3] observers tend to exaggerate the effect of such catastrophes on the world economy. The virtual elimination of an American city by Hurricane Katrina, for example, hardly moved the markets. Chernobyl, a non-event, was depicted as a disaster chiefly by the opponents of nuclear power. A 9.0-magnitude earthquake and tsunami in Japan killed 27,000 people and inflicted $100 billion in damage while the nuclear plant in the middle of the calamity inflicted virtually no damage on anyone, yet Taleb still grouses about the threat of the nuclear plant. The 1987 stock market crash, much cited by Taleb as a black swan, was a trivial event (as I told the press the next day), and it was mostly corrected within six weeks. Human beings are not turkeys, and we take consciously calculated risks in full awareness of the possibility that they may not pan out. We might even choose to go hunting with Dick Cheney without any illusions. We know all too well that the world contains plenty of axe-wielding madmen or drunken drivers or terrorists who may ruin our day in the sun.

Taleb extends his theory of stock-picking (which is, briefly, that with a large enough cohort of pickers, some poor pickers will receive acclaim as prophets through mere luck) to companies "that engage in volatile businesses." Together with survivor bias, volatility means some clinkers will seem to excel. A cohort of ten thousand companies that engage in volatile businesses will eventually dwindle to 1,250 "stars"—all perhaps attributable to pure luck. Might they include Applied Materials, Novellus, Lam Research, and ASM, the producers of the complex and exquisitely calibrated machines that manufacture microchips? Who knows? To some imponderable degree, Taleb and his followers attribute most outcomes to randomness and luck. He castigates us turkeys, who whistle blithely past "the cemeteries of silent evidence"—where the non-surviving companies or stocks or whistling turkeys are interred—congratulating ourselves on our survival instincts, business acumen, or stock–picking skills.

Since real-world outcomes scarcely differ from a random pattern, Taleb concludes that the results are random rather than planned. But his charts merely map out statistical patterns. Taleb believes that these statistical patterns show that the world is chaotic and unpredictable, that expertise is mostly delusional, and that anyone who argues to the contrary has fallen for the narrative or ludic fallacy, "fooled by randomness."

I believe, however, that the fool of randomness is Taleb. Virtually any sequence of data points, if extended far enough by a computer program, can be rendered a random wash or walk. But it is not randomness but real creativity and creation that matter to the world and to investors.

Inventions like the microchip or the laser; or the antibiotic, entrepreneur-ial initiatives like Google, Applied Materials, and Wal-Mart; or the mil-lions of small-business innovations that animate every community—these are the creative acts that enable human beings to survive and triumph. To judge the sources of a creative pattern, a chart of its ups and downs is nearly irrelevant.

On the cosmic scale, survivorship bias explains the *anthropic prin-ciple* in science, whereby the improbability of human life is attributed to the presumed existence of zillions of failed experiments or parallel uni-verses that preceded the one whose success we now enjoy, and is subject to indices of probability. Taleb invokes this survivorship principle to discredit the idea that improbability is evidence of Providence.

Taleb falls for the two great temptations of the twenty-first-century intellectual disappointed by the failure of twentieth-century science to consummate its great project of rational determinism. In physics, math-ematics, cosmology, and psychology, reason collided at every turn with an insuperable logical barrier of incompleteness, uncertainty, paradox, incomputability, or recursive futility: Kurt Gödel's incompleteness proof, showing that all logical systems are dependent on axioms beyond them-selves; Werner Heisenberg's uncertainty principle, showing the limita-tions of all measurement in physics; the aporia of Einstein's quest for a Grand Unified Theory, showing even physics to be irreconcilably divided and inconclusive; the ever-proliferating entities of the Standard Model of particle theory, further deferring the dreams of cosmic unity; and the continual shocks emerging from the computer science of Alan Turing, Emil Post, Alonzo Church, Gregory Chaitin, and John Searle, demon-strating the acute and fundamental limits of computational systems, mathematical computability, and computation as an analogy for the mind and consciousness.

Faced with this collapse of the twentieth-century agenda of scientism, twenty-first-century intellectuals have looked to information theory for a solution. They have taken solace in "randomness," in "chaos," in fractal Rorschachs, in neuro-scientific behaviorism, even in Darwinian "survival-of-the-fittest" genes and information "memes" as a kind of "universal solvent" of all other scientific explanations.

These intellectuals misunderstand the meaning of information theory. It does not afford new ways of belittling the creativity of human beings; it enhances and ensures their freedom. The most important pioneer of information theory was Kurt Gödel, who in 1926 explained the futility of all the schemes of hermetic and determinist logic that entranced intellectuals during the last century. Mathematics and the physical systems based on its logic cannot be complete. They necessarily rely on axioms beyond the logical system and irreducible to it. Gödel's breakthrough found confirmation from the computer theorists Alan Turing, Alonzo Church, and Emil Post, the chemist-philosopher Michael Polanyi, the mathematicians David Berlinski and Gregory Chaitin, and the philosopher John Searle. All their findings projected human mind beyond the compass of determinist machines.[4] They exalted the conscious human agent above a mechanistic regime. The materialist superstition was overthrown.

Beyond providing the foundation for entrepreneurial creativity, the information theory developed by Claude Shannon and his associates defined information entropy as surprise, a necessarily subjective concept, even if measured by an objective mathematical process. Perhaps the simplest and most suggestive definition of entropy is as a measure of *freedom of choice*: the higher the entropy, the larger the bandwidth or range of selection. A high-entropy economy is necessarily a free economy.

Within a free economy, entrepreneurial creations will appear to be random, whether measured by their earnings or profits or stock value. Many analysts follow Taleb in concluding that, in general, capitalist outcomes are devalued by this randomness. Wall Street appears to follow a random walk. It is impossible to outperform the market because its randomness renders it devoid of further information. The record of prices is compressed as fully as it can be and thus it cannot be outguessed. Taleb, Keynes, and many sophomoric followers have therefore dismissed the market as essentially a casino.

Information theory yields a different conclusion. A casino is indeed a fully transparent system with no additional information available. All players know all possible information. Assuming a fair roll of the dice, the odds and the payout are the only relevant data. Because the results are random, the long-term outcome is fully predictable. A series of creative surprises in the market, viewed from the outside, may be indistinguishable from a series of random events. But while random events convey no information, each creative surprise reflects a long and complex process of invention, planning, production, and marketing. In other words, it is important to observe these creative surprises not through an oscilloscope, which only registers the statistical ups and downs, but through a microscope, which reveals the details that the statistics (earnings data, for example) obscure and that make each creative surprise unique. To declare that the apparently chance oscillations of stock prices or of corporate performance signify the unpredictability of results is truly to be fooled by randomness.

Claude Shannon was certainly not fooled by randomness. Discussing the stock market in 1987, a portentous year for stocks that saw the robotic runaway crash in October, Shannon said:

> We do study the graphs and the charts. The bottom line is that the mathematics is not as important in my opinion as the people and the product. A lot of people look at the stock price, when they should be looking at the basic company and its earnings. There are many problems concerned with the prediction of stochastic processes, for example the earnings of companies. When we consider a new investment, we look carefully at the earnings of the company, and think a lot about the future prospects of the product. We're fundamentalists, not technicians.

"Are you lucky?" asked his interviewer.

> Far beyond any reasonable expectations. You know economists talk about the efficient market, and say everything is equalized out and nobody can really make any money, it's all luck and so on. I don't believe that's true at all. These are our current stocks, some of which we have only held a short time. The annual growth rates are punched out by our machine

there every night, a prehistoric Apple II which Steve Jobs wired together himself. The annual compounded growth rates of these stocks since we bought them, most of them quite a few years ago, are 31 percent a year, 11 percent, 185 percent (that one we haven't had too long), 30 percent, 31 percent, 181 percent, 10 percent 18 percent, 114 percent, 21 percent 2 percent, and 27 percent. (laughs) That's the full list of our holdings.[5]

I have to confess that I originally wrote off Taleb as a rather shallow provocateur and best-selling entertainer. But Taleb's next book, *Antifragile*,[6] is a high-entropy marvel that avoids most of the errors in *The Black Swan* and presents a fascinating and original new version of information theory. He even acknowledges that people like Shannon, whom he shows no signs of having read, are not merely lucky when they far outperform the stock market.

On the surface, *Antifragile* addresses different matters than Shannon does and decorates them with literary flourishes alien to the MIT scholar. But in essence, Taleb's "antifragility" expounds the same underlying theory as Shannon's, which he elaborates and extends with dazzling virtuosity.

Taleb describes three "systems"—whether economies, businesses, or individuals. A system is "fragile," "robust," or "antifragile," depending on its response to volatility, that is, to unpredictable change or stress. Fragile systems break under volatility. Robust systems endure volatility with tolerable damage. Antifragile systems thrive and improve under volatility.

Fragile systems are optimized for efficiency. They eschew waste, redundancy, and diversity, and they insure or otherwise protect themselves against perils and stress. Planning ahead, they are dependent on prediction or detailed knowledge—focus groups, marketing surveys, and opinion polls. They exploit their core competencies, optimizing profits by eliminating cost and duplication. They streamline and outsource. They pluck low-hanging fruit. In the terms of Clayton Christensen's *Innovator's Dilemma*, they are "sustainers."[7] In good times, they prosper. But they abhor volatility, shocks, or stresses. Ultimately they are vulnerable to black swans or disruptors.

On a graph, the curve of fragility is an upside-down cup, which economists call *convex* but which Taleb, a mathematician, dubs "concave" or "negatively convex." The point is that the fragile system relies on a pattern of outcomes that moves monotonically upward, like a growing business or economy, before eventually flattening with diminishing returns, and finally plummeting from an unbidden catastrophe.

In the center of Taleb's scheme are robust systems, which are resistant to stress. They may be built with certain redundancies that allow them to survive disasters that would bring down non-redundant systems. But they are subject to deterioration. Their curve is a straight upward-trending line showing a downward bias with time and stress.

On the right are Taleb's antifragile systems. Not only can they survive stress like a robust entity, but they also are actually strengthened by shocks. Not only can they endure volatility, but they also flourish in a tempest. They are not so much redundant (duplicative) as what we might call "eudundant" (revealing new and better capabilities under challenge). They display hormesis (strengthening with stress) and go beyond it in a spiral of gains. Taleb's mathematical description of the antifragile curve is convex. It is a right-side-up cup with an exponential upside. Its black swans are favorable.

The antifragile strategy is what Taleb calls *optionality*, or exposure to diverse outcomes; tinkering and *bricolage*; empiricism and opportunism; accepting the downside of many small ventures in order to be open to the positive black swan of exponential gains. It is the strategy favored by Shannon.

The very nature of creativity is that it always comes as a surprise to us. If it didn't, we could predict it and plan it and it would not be necessary—socialism would work. In a free economy, a high degree of apparent randomness does not mean actual randomness. An apparently random pattern is evidence not of purposelessness but of an entrepreneurial economy full of creative surprises.

What Taleb derides as the narrative fallacy, the ludic fallacy, the teleological temptation, or the prophetic impulse is the behavioral style that makes us human. It is the baseline by which we can identify the unusual or the innovative. Our imaginative and creative faculties allow us to escape the surrounding random jungle—or desert—and construct lush civilizations with elegant cafes and other comfortable settings for hedonic writers like Nassim Taleb. They enable us to avoid an obsessive focus on the irrationalities and dangers of life, and to survive to exaggerate our responsibility for our achievements.

The Outsider
Trading Scandal

EVERY JULY IN LAS VEGAS, the economist-impresario Mark Skousen runs an uproarious conference called FreedomFest, "where free minds meet to celebrate 'great books, great ideas, and great thinkers.'" You gotta go. Amid this dazzling spread of ideological delicacies—replete with debates, antics, and theatrics—even old codgers like me can have fun. My contribution to the concluding gala at the 2011 festivities was a galvanizing dance on stage with a sultry supply-side Ph.D. with long blonde tresses and exciting Laffer curves. What won't I do to promote my books?

During the previous three days, you could actually learn stuff, debating value versus vanity, technology boom versus technology bust, day trading versus buy-and-hold, costly climate change versus free weather, efficient markets versus random walk, and risk-enhanced riches versus gamblers' ruin.

A high point for many was a speech by the Princeton guru and bestselling author of *A Random Walk on Wall Street*, Burton Malkiel, who made a powerful case that the best and most objective guidance to the future movement of the markets comes from the market itself.[1] *Investor's Business Daily*, *Barron's*, and scores of investment newsletters regularly offer similar advice.

It seems plausible. Watch the movements of the market and interpret their significance. No other source of information can yield so objective a view of securities as the price changes themselves. Reflecting all the knowledge available in the market and all the decisions to buy and sell, the patterns of change in market prices are empirical, scientific data. By comparison, the information divulged by companies and experts is skewed by bias, subjectivity, self-interest, and murky interpretation.

I responded to Professor Malkiel with the contrary argument that without fundamentals—without close and necessarily subjective scrutiny of the actual performance and potential of each company—markets would contain virtually no information at all. Markets would slouch through the kind of random walk that Malkiel describes. In a random world, luck would dominate skill. You could do just as well skipping the FreedomFest investment seminars and gathering around the gaming tables and slot machines downstairs. Like casinos or lotteries, markets would come under the inexorable law of gambler's ruin.

What is the one thing that investors should know to avoid this grim fate? The economist John Mauldin wants to know. He collected the views of some of us in a volume titled *Just One Thing*.[2] My answer, reiterated at FreedomFest, was that they should know that a capitalist economy or a stock market is not a casino or a lottery. Nor is it a physical or material system. It is an arena of information, governed by mathematical laws of information similar to the laws that determine the capacity of telephone lines and wireless spectrum—the same laws of information that shape biological change through the genetic code or shepherd calls through your CDMA digital smart phone. In markets the winners are the people with the best information, mostly inside information.

Technical analysts will parade their market models down the runways. Comely regularities and curvaceous symmetries—what's not to like? A popular school of analysis celebrates the market as a fractal, displaying the fashionable swirls and whorls and "strange attractors" of chaos theory. In recent years, Benoît Mandelbrot published his *Misbehavior of Markets*,[3] persuading many that the laws of fractals and physics capture something deep about the behavior of markets.

What does information theory tell us about such ideas? Stanford University's Thomas Cover, the leading information theorist of the day, put Mandelbrot's set—the colorful filigreed whorls of intricate design and apparent complexity of Mandelbrot's fractal printout—on the jacket of his canonical book, *Elements of Information Theory*.[4] From movie posters to book covers to power point décor, graphic artists often use the Mandelbrot set as a symbol of dense information.

But information theory itself is full of surprises. Inside the jacket of his book, Cover writes, "The information content of the fractal on the cover is essentially *zero*."[5] If the measure of complexity is the number of lines in the computer code needed to produce the effect, Mandlebrot's fractal, the product of a simple computer algorithm, bears virtually no content at all. It is all froth on a core of simple algebra.

In a 2005 book, *A Different Universe*, the Nobel laureate physicist Robert Laughlin of Stanford describes the study of such froth as "baubles" and "supremely unimportant."[6] It is like analyzing water by focusing on the bubbles as it boils, a phase-change phenomenon still not understood and full of enigmas. From climate-change models to economics, such frothy data analysis is the epitome of spurious science. It focuses on trivial patterns yielding small or chaotic effects that are divorced from the actual substance of causes and consequences.

In a similar way, information theory is the nemesis of those who would reduce markets to material laws. As manifestations of the interplay of human minds, markets are more analogous to biological phenomena. As the controlling knowledge of biology resides deep inside the nuclei of cells, so the controlling knowledge of economics resides deep inside the companies that make up the market. You cannot predict the future of markets or companies by examining the fractal patterns of their previous price movements. There is no information there.

The contrary temptation has persisted throughout history. People have perennially tried to read the mind of God or the destiny of men, nations, and companies in patterns in the sands or the stars. But you cannot know a man by contemplating the shape of his head or reading the lines on his hands or examining the alignment of the stars at the moment of his birth. Similarly, you cannot predict future movements of markets by weighing the current patterns of stock prices. There simply is not enough information in current prices to reveal future prices. As the investor Ken Fisher observes, "stock prices are not serially correlated."[7]

The technical analysis of seasoned outsiders can occasionally be effective, particularly when guided by an intuitive or stealthy mastery of fundamentals. But technical analysis is essentially parasitic. It is *outside* information. For its validity, it depends on the fundamental judgments of insiders, and the insights of knowledgeable analysts who appraise the DNA of companies: their management, their financial data, and their technological endowments.

The promise of the Internet was "infocopia"—the instant spread of detailed information. Largely fulfilled, this promise has brought the world more instantaneous information on more companies and securities more cheaply than ever before. So why in the midst of an information age, when capital and data zip around the planet at the speed of light, do markets reach peaks of volatility resembling tulip auctions in 1690, when carrier pigeons were the fastest transmitters of information?

A key reason is the *outsider* trading scandal. The law for information disclosure by public companies and aspiring public companies prohibits the release of materially significant information unless it is published simultaneously to the world. This well-meaning rule is supposed to create a level playing field where no investors have the advantage of inside knowledge. But a level playing field means no information, since information is inherently unleveling. Information, like life, creation, and enterprise, brings disequilibrium. U.S. securities law manages to reduce the amount of real information in stock prices.

Less information means increased volatility and more vulnerability to outside influences. With the entire field of information about companies a regulated arena, information does not bubble up from firms spontaneously in raw and ambiguous form with executives and engineers freely expressing their views and even investing on the basis of them. It emerges as various forms of processed public relations.

Insider trading laws are intended to prevent fraud, but they do not prevent criminals from manipulating markets. Criminals, by definition, observe the law only to break it more ingeniously. The idea that you will frustrate Bernard Madoff by reducing information for everyone else is like trying to prevent airplane hijackings by making all passengers lie about what's in their luggage. The regulator feels like he's doing something, but it does not affect the criminal.

The regulation of material information instead sharply inhibits the flow of inside news from companies. Inside information—the flow of intimate details about the progress of technologies and product tests; changes of strategy, research, and development; and diurnal sales data— is in fact the only force that makes any long-term difference in stock performance. Yet it is precisely this information that is denied to public investors. Information about technology cannot trickle out day by day from different inside sources to knowledgeable people who might grasp its significance. Instead, information is parceled out by departments of public affairs and investor relations under the guidance of lawyers. The resulting press releases are mostly zero-entropy documents containing recycled marketing materials, all couched in cagey legalese and hemmed in by safe-harbor statements often longer than the release itself.

Entrepreneurial information from deep inside companies, not from the investment counsel or PR firm, is the chief real knowledge in the economy. Acquiring and comprehending it is the chief work of entrepreneurs working inside and outside their companies. Such knowledge is by no means infallible. Insiders often get it wrong. But nothing else is of much value at all. By excluding inside news from influencing the day-to-day movements of prices, the United States effectively blinds its stock markets. What should be a steady outpouring of knowledge—some of it hype; some of it confusing; most of it ambiguous, like business life itself— is reduced to a series of media events that leave out everything interesting.

In these circumstances, quarterly financial reports and merger and acquisition announcements are pivotal. Concentrated rather than diffused, this information takes the form of discrete revelations—P&L or M&A—that themselves are the targets of theft and manipulation. Thus the government's control of information creates more binary moments of disclosure and opportunities for insider trading. Moreover, because only acquiring companies can legally learn all the intimate inside details about companies they are purchasing, mergers and acquisitions become the decisive moments of value recognition. Anticipating them becomes a major preoccupation of analysts. In recent years, M&As have outnumbered IPOs in the technology field twenty to one.

With inside information banished from public markets, privateers capture the wealth. Gains go to the residual inside traders who are legally permitted to learn the intimate facts of the companies in which they invest. The big winners include conglomerateurs like Warren Buffett of Berkshire Hathaway and Jeffrey Immelt of General Electric. Berkshire Hathaway and GE are not companies at all, actually, but portfolios of diverse assets. Their strength is full access to inside knowledge about their holdings and potential purchases.

The other winners are venture capitalists like Donald Valentine of Sequoia, John Doerr of Kleiner Perkins, and, until recent years, Mitt Romney of Bain Capital. They too command full knowledge of firms they fund. When Google went public at eighty-seven dollars per share, most of the returns went to venture partnerships such as Kleiner, which had bought the shares at forty cents. The irony is that when the company went *public* its information went *private*. Submerged in the enforced secrecy of fair disclosure regulations, Google drastically reduced the flow of information to its shareholders. The public was left largely in the dark, learning little about Google's holdings outside of quarterly announcements, occasional press releases about the company's charities and solar dabbling, and business magazine personality profiles of Larry Page, Sergey Brin, and Eric Schmidt.

Such markets are vulnerable to outside information—the outsider trading scandal. Working in the dark, investors become paranoid and jump at every movement in the shadows. They debate the implicit punctuation in speeches by Ben Bernanke. They weigh merger rumors to the milligram. They even speculate on what will be the next company deemed to hold an "ascendant technology" by the Gilder Forum. They contemplate technical charts and invest on the basis of momentum—a method that is by definition for the ignorant.

Benighted investors become manic-depressives. They fall prey to pundits and politicians who may know even less than they do, but who command the media. With market prices moving by multiples on the basis of political noise from the Federal Reserve or the Treasury, investors develop hair-trigger reflexes. When the Fed makes a mistake on interest rates, when the government blunders again on broadband policy or immigration rules or tax rates, when a cabinet secretary talks down the dollar, the markets overreact. Volatility is an effect of the very ignorance that the new information tools are designed to overcome.

Even on the World Wide Web, blinded pundits cover blinded markets. Analysts who are fenced off from any "material" information not divulged at once to the world focus on the personalities of executives and on financial data, considerations that are necessarily retrospective and thus irrelevant to future prices. Because of information-disclosure laws, all data must be closely held until they are officially announced. No facts can be revealed or verified until they are fully understood by firms' executives and deemed safe to be divulged without risk of future embarrassment. Such news is obsolete by the time it is announced. The *omertà* that governs corporate management and the paralysis of insiders around them leaves markets at the mercy of misinformation, momentum, and so-called "technical" twaddle about the market itself.

Because these distortions are amplified and propagated under the supervision of the Securities and Exchange Commission, the huge expansion of financial news coverage has not produced more rational and informed pricing of stocks. The essence of federal securities laws and regulations is "Don't invest in anything you know about." State governments, reasonably enough, respond by inviting their citizens to invest in the state lottery, "where no one knows more than you." Sanctimonious pundits recommend collections of stocks covering the entire market and containing no information at all. John Bogle of Vanguard has become a media hero for stultifying markets through his index funds. Executives at companies avoid insider-trading rules by putting their own purchases and sales of stock on an automated program. This shields them from litigation and the public from the information that insider activity might otherwise impart. As prices contain less information, the market becomes a more perilous arena.

As Vigilante and Redleaf put it in *Panic*, "In pursuit of fairness, the SEC is trying to create a market in which all available information circulates perfectly." Perfect competition is the implicit ideal. "No doubt they would prefer this to be a market of perfect information (which is impossible). But since they can't do that, they are creating a market of scarce information, in which the job of resolving uncertainty contains a much greater share of luck than of judgment. It's the replacement of judgment with guessing, entrepreneurship with luck, that is the general problem of which technical trading is only an instance."[8]

To realize the benefits that the Internet could bestow on those information markets that focus on stocks, the current "fair disclosure" rules should be rescinded. At a minimum they are in violation of the First Amendment. What can be the meaning of freedom of speech under a regime in which criminal penalties are attached to the pillow talk of an IBM executive in the arms of a siren or to passing comments in the dentist's chair by a vice president of a takeover target? Fraudulent manipulation of shares can remain prosecutable without damming up the entire flow of corporate information.

Information wants to spread freely, and the more information that is incorporated in stock prices, the more robust the market can be and the less subject to manipulation, euphoria, and panic. Through the Internet, stock exchanges could cease to be Keynesian casinos and fulfill their real role in the intelligent investment of capital. Greed and fear can only give way to knowledge if knowledge is legal.

The Internet is already full of sources of competitive analysis and technological expertise. As new companies emerge, there are powerful ways of obtaining and revealing the troves of inside knowledge that determine their destiny. The builders of some of these new "knowledge networks" have been indicted for their trouble in the SEC's insider-trading witch hunts. But others, such as the Gerson Lehrman Group, have prospered with software that provides investors with quick and legal links to hundreds of thousands of available experts. A new generation of information companies, focused on the real sources of value in markets, is creating a new topology of information amid the leveled playing fields of government-enforced ignorance. Watch for these developments and take heed. Skill and information are your remedies against the dismal economics of gambler's ruin. The fundamentals will ultimately prevail. Inside knowledge is harder to get under the current rules, but it remains irrepressible.

The Explosive
Elasticities
of Freedom

THE LAST TIME I saw Peter Drucker was early in the year 2000. He was keynoting a *Forbes* conference in Seattle for CEOs, and they had wheeled the great man out to the middle of the stage in a fluffy easy chair. Just over ninety years old, Drucker was a Delphic presence at the conference. Everyone leaned forward to hear what he had to say.

Suddenly a collective gasp rose from the assembled CEOs. The conference managers stood stricken. "For the love of Malcolm's motorcycle," we wondered, "What is this?"

The hoary sage's balding pate had flopped back in the chair as if he had fallen asleep ... or worse. I was horrified. Perhaps *Forbes* had erred in staking the success of a major conference on an aging guru seemingly in parlous health.

Then his entire body crumpled forward. I was ready to run up to catch him if he should tumble toward the crowd. But he somehow caught himself. His eyes opened, and he looked out intently at the throng of CEOs. Everyone sighed with relief. He was awake. He had their attention.

Pointing his right forefinger toward the audience, Drucker growled, "I have just one thing to tell you today. Just one thing...."

"Wow," I said to myself, "it better be good."

"No one," he continued, "but *no one* in your company knows less about your business than your *see eff oh*."

Huh?

This was the era of the heroic chief financial officer (CFO). Scott Sullivan of WorldCom, Andy Fastow of Enron—clever, inventive folk like that. You remember them. Across the country, CFOs were in the saddle. GE's CFO, Dennis Dammerman, was given heavy credit by Jack Welch for the ascent of General Electric and its transformation into a leviathan of finance. CEOs would not move without consulting them.

What could Drucker have meant?

He was stating Law Number One of enterprise: Knowledge, particularly financial knowledge, is about the past. Entrepreneurship is about the future.

CFOs deal with past numbers, with *accounts*. By the time they get them all parsed and pinned down, the numbers are often wrong. In effect, CFOs are trying to steer companies by peering into the rearview mirror. Past numbers have little to do with future numbers. Remember Ken Fisher's principle: company returns and stock prices are not serially correlated.

As Drucker points out, company returns are not determined within companies. In companies are no profit centers, just cost centers. Whether a particular cost yields a profit is the result of decisions made outside the company, by customers and investors. Reaching customers and investors takes outward-looking vision and leadership, not inward-looking problem solving.[1]

Once again, we return to Drucker's rule: entrepreneurs should pursue opportunities, not solve problems. Solving problems sounds good, and it is the CFO's specialty—and pitfall. You end up staring into the past. You feed your failures, starve your strengths, and achieve costly mediocrity. Instead, pursue opportunities.

Drucker's view of CFOs was relevant far beyond the boundaries of the individual companies that hire them. CFOs in effect are "economists" studying the economy of one company. They abstract from existing data to create an orderly picture of the company at a given moment, focusing on apparent sources of profit and loss. The CFO makes sure his firm has enough low-entropy cash in case things go badly. He makes sure the company can sustain reversals. Fastow, Sullivan, and the others were disasters because the last thing you want from the finance department is bravado. The daring young man on the flying trapeze makes a terrible CFO.

The reason economists cannot explain growth is that economists are just macro-CFOs. When they do their jobs well they can warn us against certain dangers, such as the crippling buildup of federal and state debt. But when economists become creative; when they seek to surprise us with their ingenuity at conjuring growth by manipulating government spending, monetary policy, and mandates; they do to the economy what Fastow did to Enron.

The advocates of growth now face a David Stockman moment. As director of the Office of Management and Budget, Stockman was in effect President Reagan's CFO. Young and bright, he was assigned the Herculean task of retrenching government spending in the midst of the Cold War. In this role, he made the covers of all the most fashionable magazines as "Mack the Knife," or "the Beltway Butcher"—an alleged mad slasher of compassionate government, ruthlessly cutting back the crucial programs that sustain the poor.

Stockman ended up capitulating to his critics. He obsequiously abandoned supply-side economics as a naïve mistake and skulked off to the dreaded private sector. Starting with the fashionable plutocrats of Salomon Brothers, he moved on to the Beltway-savvy Blackstone Group. Commendably rebelling against Blackstone's cronyism, he left to make a bold entrepreneurial bet of his own on the revival of the Rust Belt in the Midwest. His Heartland Industrial Partners was an admirably contrarian failure, raising $1.35 billion and losing most of it. That can happen in enterprise.

Stockman apparently learned the wrong lesson. Echoing the economic wisdom of President Obama and Paul Krugman, he reemerged during the 2012 election season as the budgetary backstabber and the nemesis of bankers. He wrote in the *New York Times*, "[Paul] Ryan's sonorous campaign rhetoric about shrinking Big Government and giving tax cuts to 'job creators' (read: the top 2 percent) will do nothing to reverse the nation's economic decline and arrest its fiscal collapse."[2] Then *Newsweek*, in its death throes, published an excerpt from Stockman's forthcoming book, *The Great Deformation: How Crony Capitalism Corrupts Free Markets and Democracy*.[3] It was a canny, captious post-hoc critique of the allegedly predatory ways of Bain Capital under Mitt Romney. A Blackstone rival far less entangled with crony capitalists and more successful, Bain Capital grated on Stockman, who complained about the "utter unfairness of windfall riches" for the "one percent."

Citing his own seventeen years of experience doing leveraged buyouts, Stockman homed in on several of Bain's investments that shockingly failed long after Romney left. He sneered at the use of junk bonds because of their association with the falsely-accused Michael Milken, and he pointed to other Bain projects that used debt in the process of restructuring. Bain's stunning overall success, according to Stockman, was mostly attributable to loose monetary policy. As Alan Reynolds observed in a superbly pungent *National Review* essay, "Stockman's ... message to Bain Capital is 'You didn't make that.'"[4]

The *Newsweek* excerpt won Stockman the expected media applause, and delighted the Democratic campaign. Yet in a splendid interview with *Reason* magazine's Nick Gillespie[5] and elsewhere in *The Great Deformation*, Stockman makes it clear that he got it mostly wrong during the campaign. Who knows what he thought he might accomplish for his libertarian faith by helping to reelect a socialist president whose only compromise with private markets has been a series of crony capitalist orgies? Stockman's book demonstrates that he disagrees with virtually every detail of Obama's and Krugman's economics. That's great, but it doesn't matter because Stockman is still chiefly a CFO totaling up America's liabilities.

Everyone knows about America's liabilities. Everyone knows that they will have to be addressed, and many know that Paul Ryan's plan would have addressed them. But the opportunity for growth-oriented American leaders today has little to do with balancing the budget or directly suppressing the nation's liabilities. It is to transcend all these problems by reversing the devaluation of America's human and capital assets, which is what renders the liabilities increasingly unsupportable. Capital is not merely a flow of power; it is an accumulation of specific knowledge.

Seventy percent of federal spending—some $900 billion—devalues human assets by suppressing the learning process that results in productive citizens. The government is paying people to be unemployed, unmarried, retired, sick, poor, homeless, hapless, disabled, indebted, drugged or imprisoned. These benefits and guarantees amount to a huge tax on escaping from dependency, rendering most available jobs, opportunities, marriages, and careers far less remunerative than promiscuous sloth. The unraveling of family life in the inner cities of America has created a welfare state for women and children and a police state for the boys. The third of young ghetto males who are in prison or on probation or on the lam are grim testimony that, as a rule, women alone cannot raise boys. But the welfare state relentlessly punishes marriage and family among the poor. As Phyllis Schlafly points out, dozens of programs designed to counteract a so-called singles penalty in the tax code have created a massive marriage penalty instead.

These problem-solving programs accomplish nothing but their own expansion. They spread dependency. They destroy lives, families, and careers by the millions. And the tax they impose on those who try to escape from dependency far exceeds that faced by the richest Americans. The "callous" ones are not those who would cut back these ruinous programs but those who would condemn millions of their fellow citizens to perpetual dependency. Reforming the welfare state is all upside, removing a crippling burden from the nation's economy and an obstacle to the pursuit of happiness.

Upside policy changes can turn around the entire American economy. They can redeem the stocks and bonds and hopes for initial public offerings dashed and devalued by mazes of promiscuous tolls and regulations. They can restore the land wasted and ruined by ethanol and windmills and druidical sun henges and water rules. They can encourage industrial innovation and venture capital. They can release the real energy resources capped and crimped by regulations, restrictions, and litigation. They can reclaim the real estate wasted and plundered by federal finance blight and insurance scams. They can renew the youthful aspirations and talents depleted by debt assumed to finance phony educations at schools of self-esteem. And they can rebuild the banks debauched by crony capitalism: federal bailouts, zero interest rates, treasury privileges, and social causes. All upside also, though Stockman will never see it, would be restoring America's military leadership, deterrence, and innovation, which Reagan and the Bushes fostered and which Obama has let wither.

Stockman may have held Ronald Reagan in disdain, but Reagan understood the Drucker rule far better. Reagan pursued opportunities while Stockman tallied up the problems. Reagan horrified David Stockman by raising government spending far more than his failed predecessor, Jimmy Carter, did. Reagan pushed the accounts of government into the red and allowed U.S. imports to surge in comparison to exports. He understood the strategic position of the United States well enough to grasp that the quality of our technology and the contributions of unleashed entrepreneurs dwarfed in importance an increase in debt.

Adverse numbers happen when you pursue the opportunity of leading the West to victory in the Cold War. Reagan's nearly trillion-dollar creative bulge in defense spending transformed the global balance of power in favor of capitalism. His tax-rate reductions and regulatory retrenchments enabled an entrepreneurial revival in the United States. Reagan's policies spurred stock market, energy, venture capital, real estate, and employment booms, revaluing America's private-sector assets upward by some $17 trillion, dwarfing the trillion-dollar rise in public-sector deficits and creating a net 45 million new jobs at ever rising wages and salaries. He even made a Social Security deal with the Democratic speaker Tip O'Neill that reduced the discounted present value of total transfer payment liabilities by six trillion dollars, making possible two more decades of Social Security solvency.

The huge improvement in the U.S. asset position, ultimately raising private-sector net worth by another $60 trillion over twenty years, did not halt until 2010. Under the Obama administration, for the first time since the 1970s, the American economy began suffering capital flight and, for the first time ever, experiencing a net emigration of high technology talent, with entrepreneurial engineers returning to India, China, and Israel.

Socialism cannot be remedied with mere accounting adjustments. The huge opportunity today is an all-upside campaign to reverse these crippling trends of government overreach and redistribution.

In the view of Stockman and other CFO-minded analysts, an outsized share of the yield of U.S. enterprise flows to capitalists and investors. Capital and labor, however, are not competitive but complementary. As workers become more productive, employers hire more, not fewer. The Reagan productivity boom propelled a huge employment boom. Capital linked with private-sector knowledge releases creativity and stimulates new employment. Drastically lower marginal tax rates on income and capital formation could endow millions of more jobs at higher pay. Replacing America's corporate tax rate—the world's highest—with an 8.5-percent business consumption tax, as Paul Ryan has proposed, would remove the incentive to avoid taxes and profits and would encourage creativity and entrepreneurship.

To the CFOs scrutinizing America's books, healthcare is a huge cost center depleting the country's resources, and Medicare is a crisis. But healthcare is not really a problem; it is a huge opportunity. The opportunity is obscured by government bureaucracy, price controls, and regulatory glut. With its 16,000 new IRS agents and no new doctors, Obamacare also restricts new medical instruments with a 3.2-percent tax on gross receipts that can capture more than 100 percent of profits. The Food and Drug Administration's rules are particularly hostile to health care innovation, adding a billion-dollar toll over seven years to the launch of a new pharmaceutical. This federal stricture pushes drug companies to focus on sex and lifestyle pills with huge markets rather than on new drugs that can keep people alive and working.

Skilled, experienced, and enjoying improving health, seniors do not have to be a burden or a problem. They can increasingly remain in the workforce as assets rather than becoming liabilities for their diminishing numbers of grandchildren. Saving Social Security and Medicare is an opportunity for keeping seniors hale and in the workforce rather than driving them out by putting punitive tax rates on their earnings and halting innovation in health care with a government takeover.

Perhaps the most obvious rule of public policy is that people will abuse any free good. Evoking unbounded demand while choking off supply, free goods and free services destroy information and lead to corrupt decision-making. In the perverse feedback loops of free goods, free health care comes to mean hypochondria and needless illness caused by needless exams and treatments, queues for an ever expanding political portfolio of mediocre services, and—at the end of the line—euthanasia under government bureaucracy. Free drugs lead to widespread addiction to existing medications and an end to medical innovation. Free money, manifested in the zero-interest-rate policy of the Federal Reserve, diverts the wealth of savers to favored governments and crony capitalists while creating shortages for everyone else.

As has been shown throughout history, a change in policy on the supply side can effect an instant and sharp improvement in the value of all entrepreneurial assets. Supply-side reforms today would launch us into a new American century.

More than two decades ago, in *The Rise and Decline of Nations,* Mancur Olson described how democratic economies gradually become sclerotic under the accumulation of bureaucracy and interest groups.[6] He saw both the decline of America and the rise of China as examples of this unfolding process. The Old Testament law commanded the Israelites to remit one another's debts every seven years. It also ordained a Jubilee year of liberty every fifty years. What we need today is a new jubilee: the remission of superfluous laws, regulations, special-interest subsidies, tax rules, and privileges that undermine the epistemic functions of a free economy, suppressing knowledge and stultifying enterprise.[7]

The CFOs of American decline disagree on many subjects, from climate change to income distribution. But they all agree that debt and deficits are the paramount threat to America's future. Politicians and opinion leaders of both parties see the control of debt as the principal point of leverage in reversing American decline. From the liberals Thomas Friedman and Michael Mandelbaum in *That Used to Be Us,*[8] who identify chronic deficits as one of the four biggest threats to America; to

Admiral Michael Mullen, chairman of the Joint Chiefs of Staff, who deems the deficit to be the chief threat to U.S. security; to the father-and-son libertarians Ron and Rand Paul; people from all sides of the political establishment share the conviction that our economic and national security problems are the result of an imbalance in the federal government's revenues and expenditures. The crisis of our time is a crisis of the national accounts.

From this standpoint, the remedy is obvious: drastic retrenchment of current spending or ruthless extraction of more revenues through tax increases—or both. Since either remedy is unreachable, this diagnosis of our ills leads to a psychological despair as acute as the economic depression. It makes Republicans like David Stockman contemplate suicidal cutbacks in national defense spending. It pushes Democrats to attack the sources of wealth with tax increases that actually shrink the government's revenue.

This diagnosis of our woes exonerates liberalism. It implies that our economic doldrums are the result not of *current* policy but of decades of bipartisan bingeing on everything from defense to Medicare to housing subsidies. Even Obama's runaway stimulus programs only recapitulated Bush's prodigal TARP. Deficit panic impels politicians into frenzies of poll-driven demagoguery, calling for a balanced-budget amendment to the Constitution and other mischief that would rectify nothing.

History tells us that the threat to prosperity is not debt but socialism. The key error of socialism is separating knowledge from power—concentrating power at the foggy peaks of government rather than distributing it through the high-entropy channels of a free economy. Socialism attempts to guarantee the value of things but not the ownership of them, and nothing destroys value more effectively than the erosion of ownership and responsibility.

Nothing is more advantageous to the socialists in our midst than a futile and confusing preoccupation with debt. Thomas Babington Macaulay provides some useful historical perspective in his discussion of Britain's fiscal straits at the end of the twelve-year War of the Spanish Succession, which had engulfed Western Europe and spilled over to North America. "When the great contest with Louis the Fourteenth was finally terminated [in 1713] by the Peace of Utrecht," he wrote in his *History of England*,[9] "the nation owed about fifty millions; and that debt was considered, not merely by the rude multitude, not merely by fox-hunting squires and coffee-house orators, but by acute and profound thinkers, as an incumbrance which would permanently cripple the body politic...." Over the next hundred years, while fighting three more fabulously costly wars, including the American Revolution,

> The beggared, the bankrupt society not only proved able to meet all its obligations, but, while meeting these obligations, grew richer and richer so fast that the growth could almost be discerned by the eye....Meanwhile taxation was almost constantly becoming lighter and lighter; yet still the Exchequer was full.
>
> ... Those who uttered and those who believed that long succession of confident predictions ... erroneously imagined that there was an exact analogy between the case of an individual who is in debt, and the case of a society which is in debt to part of itself; and this analogy led them into endless mistakes.... They were under an error not less serious, touching the resources of the country. They made no allowance for the effect produced by ... the incessant effort of every man to get on in life. They saw that the debt grew, and they forgot that other things grew as well as the debt.[10]

Anyone who imagines that the British national debt during this period was smaller as a proportion of GDP than the government debts of the current era should think again. The debt that all the experts of the day predicted would ruin the British economy reached a pinnacle of more than 250 percent of GDP in the 1820s, when the British Empire was flourishing. Until around 1865 it did not drop below 100 percent, which is where America's debt stands today.

It is perverse and self-defeating for conservative CFOs to focus on the federal government's profit and loss statement as if it were some test of fiscal virtue and the path to national revival. The key issue is not the prodigal spending but the corruption of the economy's signaling systems by the public-sector noise of bribery and manipulation.

Stockman is right on target about the menace of crony capitalism. But he is so befogged by his focus on raw numbers rather than on information and value that he cannot tell the difference between Reagan and Obama. The real threats to the United States are the forces of decline that Mancur Olson identified—parasitical litigiousness, single-issue movements, entrenched bureaucracies, and devotion to the privileged past by poll-driven politicians and their cronies.

Crony capitalism feeds on campaign finance laws that give union and corporate political action committees the power of unlimited spending while individual citizens are restricted to $2,500 per campaign. The famous "occupiers" of Wall Street actually got this right. Corporations and unions are not aggregate citizens or political entities and should not be direct players in politics. But individual citizens should be allowed to spend whatever they wish. Our campaign finance system delivers our politics into the hands of monomaniacs, which is what PACs are. Unlike a citizen, who has a range of interests and a stake in his community, a PAC is by definition devoted to a single cause. Its donations, therefore, are indistinguishable from bribes.

Economic renewal requires stable, predictable public leadership without wild shifts and manipulative surprises. The government recently erected an almost insuperable barrier to growth when it granted the EPA arbitrary power to regulate carbon dioxide across the entire economy. All new energy enterprises now face a long series of ambushes by the public sector, which ought to be a responsible and predictable carrier for the upside surprises of private-sector creativity.

A transformation of policy on the supply side would effect an instant and sharp improvement in the value of all entrepreneurial assets. A rise in the worth of creative companies would spark an upsurge of investment in innovative projects, an appreciation of related real estate, and a reorientation of education toward real innovation. We would see job expansion, increased investment in research and development, high tech immigration, and investment and venture capital outlays similar to what we experienced under Ronald Reagan and his successors from both parties.

Or is all that just pie in the sky? Economists are usually pessimists. They believe in the intractability of existing conditions, whether the inflation of the 1970s or the deficits of today. But here again, a glance at history is instructive. Massive improvements have in fact been quickly achieved with a reversal of policy. After World War II, when ten million demobilized servicemen returned to an economy that had to be converted from a garrison state to civilian needs, economists steeled themselves for a renewed depression. A sweeping Republican victory in the Congressional election of 1946, however, brought an end to the wartime government-planning regime. Dropping from 42 percent of GDP to 14 percent, government spending plummeted by a total of 61 percent between 1945 and 1947. One hundred fifty thousand government regulators were laid off, along with perhaps a million other civilian employees of government. The War Production Board, the War Labor Board, and the Office of Price Administration were dismantled.

Every Keynesian economist confidently predicted doom. Sounding exactly like his future student Paul Krugman, who would beg Obama for trillions in additional "stimulus" spending, Paul Samuelson in 1945 prophesied "the greatest period of unemployment and dislocation which any economy has ever faced." Arnold Kling of the Cato Institute has observed that "as a percentage of GDP the decrease in government purchases was larger than would result from the total elimination of government today."[11]

As Paul Krugman points out, nominal GDP, as measured by economists, did drop a record 20.6 percent in 1946 when government spending plummeted. But a drop in government spending after a war does not depress private creativity; it unleashes it. Judging the public-sector contribution by its cost is the great error of Keynesian economics. With the rearview perspective of a CFO contemplating last year's data, economists pondering the statistics of 1946 missed everything that was going on. GDP cannot capture the reality of an economic transformation because it necessarily misses the drastic changes in the noosphere—the subjective realm of ideas and attitudes, goals and inspirations that determine the worth of all spending in the economy.

A revaluation of all private-sector assets was underway, brought about by the removal of public-sector burdens that were previously crowding out private ventures. A flow-based concept, GDP does not take into account positive or negative changes in a country's balance sheet as shaped by the noosphere.[12] In fact, here was no depression. Instead, the Great Depression, which had continued through the war disguised by price controls and necessary defense spending, at last came to an end.[13] Economic growth surged by 10 percent over two years and the civilian labor force expanded by seven million workers. Unhindered by wartime controls, the private sector proved eminently capable of adapting to the new opportunities, launching a ten-year boom despite self-defeating tax rates on investors as high as 91 percent. The Republican Congress compensated for the high rates by introducing joint returns, effectively cutting taxes in half for intact families. Corporate taxes dropped drastically, and the tax burden, measured by government spending, fell more dramatically than at any other time in American history. Low inflation and privatization led to a resurgence of large manufacturing corporations that had been honed into lean and hungry productivity engines during the war.

New Zealand enjoyed a similar turnaround in the mid-1980s. Once among the world's richest countries, with flourishing trade in agriculture and building materials, New Zealand had become mired in socialist stagnation. A series of compliant conservative governments that still believed in the popularity of socialist oppression and the impossibility of dismantling it without peril had left the country unable even to feed itself after twenty years of massive agricultural subsidies and supports. With a top income tax rate of 66 percent and a government share of GDP of 45 percent, New Zealand's economy was paralyzed.

The saving change in policy came, surprisingly enough, with the election of a Labor government, which adopted a policy of zero-base budgeting for all government departments. New Zealand sold off its nationalized airlines, railways, airports, seaports, bus lines, banks, hotels, insurance firms, maritime insurance companies, radio spectrum, printing facilities, forests, irrigation schemes, and an array of other holdings. It abolished the farm programs that in 1985 were supplying 45 percent of all agricultural income.

Maurice McTigue, one of the architects of the transformation, reports the results: "A decade later, New Zealand had one of the most competitive economies in the developed world. The government share of GDP had fallen to 27 percent, unemployment was a healthy 3 percent, and the top tax rate was 30 percent." Taxes on capital gains, inheritances, and luxuries were eliminated, and excise duties were removed. Revenues surged and "the government went from 23 years of deficits to 17 years of surpluses and repaid most of the nation's debt"—chiefly through record economic growth that restored New Zealand as one of the world's richest nations.[14]

Perhaps the most impressive results came in agriculture, which underwent the most abrupt change in policy and the cold-turkey withdrawal of subsidies. From its mendicant status as a large importer of food, New Zealand was transformed into one of the world's most creative and profitable food exporters. Its dairy output multiplied from just a handful of products—basic milk, cheese, and butter—to some 17,000 variations, and its success at exporting cheese and butter products was great enough to provoke charges of unfair trading practices from the state of Wisconsin.

During the same period, Israel made a similar transition from a socialist economy in crisis to privatized and deregulated free enterprise. Allegedly complicating the Israeli challenge was the arrival of close to a million Russian Jews, many of them possessing advanced educational degrees and mathematical and engineering skills.

Like the United States and New Zealand, Israel benefited from a dramatic change in political leadership that signified a new attitude toward enterprise. The conservative Likud party that assumed power in 1985 under Yitzak Shamir reflected the pro-enterprise evangelism of Zev Jabotinsky and his secretary Benzion Netanyahu, the father of of the future prime minister, Benjamin Netanyahu, who was then Israel's ambassador to the United Nations. With tax rates dropping some 30 percent and public ownership of leading corporations plunging from some 80 percent to 20 percent, the economy took off.

In a decade, Israel went from an inflation-ravaged industrial cripple to a world leader in technology. Between 1991 and 2000, venture capital outlays in Israel rose sixtyfold, as private-sector investments overwhelmed an initial government subsidy program. A 2008 survey of the world's venture capitalists by Deloitte & Touche showed that in six key fields—telecom, microchips, software, biotech, medical devices, and clean tech—Israel ranked second only to the United States in absolute terms. In the vital area of innovation in water use and conservation, it decisively led the world. Gaining half its water from desalinization plants, adopting inventive forms of drip irrigation, and recycling 95 percent of its sewage, Israel actually managed to reduce its net water usage by 10 percent since 1948. Meanwhile, as other Middle Eastern countries faced acute water shortages, Israel's agricultural production rose sixteen-fold, its industrial output surged fiftyfold, and its population increased tenfold.[15]

A rapid change of policy in Canada has been similarly catalytic. In 1994, government debt was 66 percent of GDP, the economy was in a slump, the unemployment rate was near double digits, and the Canadian dollar was slipping far below its American counterpart. As in New Zealand, a leftist government took office with a promise of drastically restructuring the nation's finances. The government of Jean Chrétien adopted a zero-sum rule for all departments, requiring any minister who wished to protect one of his programs to make a larger cut in the cost of another program. Privatizing a range of enterprises, including railways, uranium mines, and air traffic control, Canada managed to drop the federal government's share of GDP from 17.5 percent to 11 percent. Though personal income tax rates were initially left unchanged and an array of loopholes were closed, the new policies were so successful that by the year 2000 the Chrétien government reduced the corporate tax rate by 7 percentage points, reduced income and capital gains tax rates, and increased the contribution limit for untaxed retirement accounts.

Under Chrétien, his Liberal successor Paul Martin, and the Conservative government that came to power in 2006, the Canadian economy experienced steady expansion, often faster than the U.S. economy, with a steady appreciation of the Canadian dollar.[16]

Surprises of resurgent growth and creativity have taken place not only in these showcase countries, but also in the United States after the crash of 1921, in Margaret Thatcher's United Kingdom, in the revival of Eastern Europe after the fall of communism, in China, and even in Sweden and Brazil. Particularly impressive is the rise of Sweden since 1993. It has cut its debt from over 70 percent of GDP in the early 1990s to close to 30 percent in 2013, while sharply cutting tax rates.

The British journal the *Spectator* tells the story:

> When Europe's finance ministers meet for a group photo, it's easy to spot the rebel—Anders Borg has a ponytail and earring. What actually marks him out, though, is how he responded to the crash. While most countries in Europe borrowed massively, Borg did not. Since becoming Sweden's finance minister, his mission has been to pare back government. His "stimulus" was a permanent tax cut. Tax-cutting Sweden had the fastest growth in Europe last year, when it also celebrated the abolition of its deficit. The recovery started just in time for the 2010 Swedish election, in which the Conservatives were re-elected for the first time in history.
>
> He continued to cut taxes and cut welfare spending to pay for it; he even cut property taxes for the rich to lure entrepreneurs back to Sweden. "There would be no Ikea without [Ingvar] Kamprad. We would not have Tetra-Pak without [Ruben] Rausing. They are probably the foremost entrepreneurs we have had in the last few decades, and both moved out of Sweden." ...If you have a high wealth tax and an inheritance tax, people emigrate because it becomes too costly to own a company. What even Borg did not expect was that his tax cut for the low-paid would increase economic growth

so much that it has almost entirely paid for itself. Borg had created ... a self-financing tax cut.[17]

All these turnarounds show that a retrenchment of interventionist government, with its subsidies, mandates, crony capitalism, and gargantuan tax bites, is entirely positive for a capitalist economy.

Bureaucrats cherish the self-serving myth that reducing government power and spending is risky. But a clear change in policy in favor of enterprise and creation can accomplish a drastic turnaround in the performance of an economy. Policy changes in the noosphere can galvanize the material economy, while attempts to stimulate the material economy directly with government spending accomplish less than nothing for economic growth. History offers a compelling answer to the suggestion that the only effective response to America's economic predicament is continued spending on favored industries; monetary bingeing for the benefit of crony banks; and reckless splurges of governmental borrowing lavished on the unemployed, provided that they do not find or create new work.

The Obama administration is subjecting the United States to the morbid subversion of the infrastructures of its economy. The public sector has become a manipulative force, aggressively intervening in the venture and financial sectors with guarantees and subventions that attract talent and corrupt it. At the same time, the government inflicts constant downside surprises of taxation, currency devaluation, and regulation, leaving only low-profit residues for real private entrepreneurs and manufacturing pioneers. We have a high-entropy carrier (government) and a consequently low-entropy message (a stagnant economy). This is the opposite of how a successful information society should be run.

These policies have caused capital losses, real estate depreciation, disinvestment, capital flight, emigration of skilled labor, and the contraction of private employment, with a near 50-percent drop in new jobs from start-up companies. U.S. policy is now so hostile to the creations of the private economy that a supply-side revolution could produce an explosive turnaround. The Keynesian pessimists would be discredited again. Just as the post-war recovery of the 1950s confounded Paul Samuelson's gloomy predictions, American entrepreneurs are waiting for their chance. And the revolution needs to begin with the tax code.

Flattening
Taxes

IN THE MIDST of the 2012 presidential campaign, the Congressional Research Service released an astonishing report. It maintained that there is no clear evidence that lowering top marginal income tax rates fosters economic growth. Tax increases, the CRS found, will spur economic growth. The supposedly non-partisan research agency known as "Congress's think tank" thus affirmed one of the most popular dogmas in academic economics.[1]

The new consensus—represented by such luminaries as Bill Clinton, Paul Krugman, and the *New York Times* economic writer David Leonhardt—seems to be that high tax rates encourage economic growth. Baffled by the upsurge of the economy after World War II, when government spending dropped by 61 percent, these Keynesian experts have identified that era's top marginal tax rate of 91 percent as the engine of the powerful postwar boom. They also look back nostalgically on the Clinton era, when a 10-percent income tax hike (and an unmentioned 30-percent cut in the capital gains tax rate as well as lower tariffs) seemed to conjure up the dot.com and telecom booms and balanced the federal budget.

The CRS report maps the numbers since 1945, showing that the postwar and 1990s surges were not anomalous—the highest personal income tax rates are often associated with higher per capita economic growth. The CRS noted, moreover, that it found no correlation between tax *cuts* and per capita growth. When pressed, the agency even disclosed some evidence that tax cuts can lead to economic stagnation.

All of this deep thinking about the menace of tax cuts evoked what might be termed discreet jubilation among the political and media elite. Leonhardt rushed to confront the Republican vice presidential candidate and tax-cut enthusiast Paul Ryan with the inconvenient data, which were arrayed in an alpine curve up across the page, dramatizing the association of tax cuts with GDP slumps. The curve climaxed in the devastating 2008 crash, five short years after Bush's famous "tax cuts for the rich." Ryan responded that "correlation does not mean causality," a statistical truism that left the *Times'* man unimpressed.

If correlation does in fact indicate causation, as Leonhardt and friends seem to insist, then we might as well investigate a connection between global warming and economic growth. In the decades after World War II, after all, the globe warmed, the economy grew, and the stock market climbed, more or less together, until 1999—the year of El Niño. Then temperatures drifted down for more than a decade, the stock market crashed, and the U.S. economy stagnated. On a graph the lines converge uncannily.

As you may have suspected, however, there are possible problems with the correlation between growth and global warming. For example, retreating a few years back in history, global warming also surged in the United States during the Great Depression of the 1930s. The temperature and economic growth lines therefore diverge in the 1930s, dashing our theory of a link between global warming and growth

The discovery that tax cuts are economic retardants and that tax hikes are stimuli was nearly miraculous. To foster growth and job creation, there would be no need to rein in taxing and spending. During the dark early days of the Reagan administration, some supply-siders, such as Arthur Laffer and me, caused a stir by claiming that marginal tax-rate reductions usually pay for themselves through faster growth of the economy and the tax base.[2] The CRS report allows us to brush off those pesky tax-cutters and return to the comfortable Keynesian proposition that government *spending* pays for itself. Government spending counteracts private-sector savings and yields growth, compounded multiplier effects, expanded employment, and confectionary reflux from an always shovel-ready Federal Reserve. The more government spends, so imply these Keynesian experts, the more revenues pour in to be spent.

This view may resemble the confidence of blowsy drinkers who insist that the more they imbibe the better they drive, but we shouldn't dismiss it out of hand. Scrutinizing more closely the numbers from the CRS, we might detect some cautionary anomalies. By 2013, marginal tax rates in the highest brackets had risen to 50 percent in many localities. A 50-percent rate gives these crucial taxpayers a greater incentive to hide income than to earn it. That 91-percent rate on dividends, for example, garnered almost no revenue because only 127 taxpayers disclosed income in this bracket or in the 86-percent bracket just below it. At a time of persistent unemployment, moreover, 50-percent rates give entrepreneurs a stronger incentive to lay off workers to control costs and lower taxes than to hire new workers to increase revenues.

In appraising the effects of tax cuts and hikes, we should look beyond the mixed results under the two Bushes and Clinton and consider also the surges after the Harding and Coolidge cuts in the 1920s, the Truman cuts after World War II (chiefly drastic corporate tax reductions and the 50-percent cut in personal rates implicit in the adoption of joint returns in 1946), the Kennedy cuts in the 1960s, and the Reagan cuts in the 1980s.

And for less parochial economists, there are hundreds of examples of tax policy to scrutinize around the globe. A 1983 study by the World Bank economist Keith Marsden,[3] extended and updated by Polyconomics in 1992,[4] compared countries with high or rising tax rates with countries with low or declining tax rates. The analysis discovered that the low-tax countries increased their government spending three times as fast because they grew their economies six times as fast as the high-tax countries did.

Over the two decades since the Marsden-Polyconomics study, some fifty countries, led by China and the former communist bloc, have sharply reduced their tax rates. Following the examples of perennial growth leaders Hong Kong and Singapore, some twenty-five countries have adopted low flat-rate tax systems. Virtually every flat tax has engendered accelerated growth and attracted increased foreign investment. Russia reduced its top individual tax rate from 30 to 13 percent in 2001 and doubled revenues over the subsequent four years. Serbia, Ukraine, Slovakia, Georgia, and Romania followed, with robust revenue responses. After adopting flat-rate tax systems, with rates eventually dipping below 20 percent, Estonia, Latvia, and Lithuania averaged 8-percent real economic growth for a decade. Poland (which flattened and lowered its corporate rates), Hungary, and Bulgaria followed with similar if less dramatic results. Of the twenty flat-tax countries, report the economists Daniel Mitchell and Chris Edwards in *Flat Tax Revolution*,[5] none has experienced unusual revenue problems, and none has rescinded its flat rates. In fact, their average personal flat rate has fallen to 16.6 percent and their average corporate flat rate has dropped to 17.9 percent, about half of the U.S. federal levels.

In the United States, meanwhile, corporate and personal tax rates have drifted upward to world-leading levels. Our taxes are higher than the average in Asia, Africa, Latin America, and Eastern Europe. Unlike the flat-tax countries, the United States has incurred world-leading deficits and debt burdens, and for the first time since the Carter years, the United States is suffering net capital flight. America has become more anti-capitalist and enjoys slower economic growth than the former communist world.

Perhaps we need not fear tax rate reductions after all.[6]

Keynesians are at a loss in explaining tax policy because they treat the economy as a system of power rather than a system of knowledge. But supply-siders, too, though on the right side of the argument, tend to confuse the issue with their stress on capitalism as a system of incentives.

The connection between lower rates and higher revenues is perhaps the most thoroughly documented yet widely denied proposition in the history of economic thought.[7] Even some of the supply-side faithful, like the estimable Bruce Bartlett, have wandered into doubt, spending much of their time chewing and re-chewing ambiguous data from the United States in the 1990s, while ignoring the tidal wave of evidence from the global tax revolution beyond our shores.

At a generation's distance, it is clear that we, the original supply-siders, bear some responsibility for the failure to persuade. It has become clear that we were not radical enough—that we allowed our own arguments to be ensnared by the mechanical economics of Smith and his heirs. Even Laffer's original and brilliant sketch, after all, functioned almost entirely in the realm of rational expectations—stimulus and response applied to poor, passive *homo economicus*. Let him keep more of the fruits of his labor, and he will labor harder, we proclaimed; increase the after-tax rewards of investment and more investment there will be.

These propositions are certainly true. People do respond to the incentives of the system. The increasingly heavy penalties on marriage and family, for example, are steadily eroding America's crucial human capital. The real argument for lower taxes, like the argument for less regulation, like the argument for capitalism itself, only begins with simple incentives for productivity.

The secret is not merely giving people an incentive to work harder or to accept more risk in order to gain a greater reward. The reason lower marginal tax rates produce more revenues than higher ones is that they release and endow the creativity of entrepreneurs, allowing them to garner more information. They can move more rapidly down the curves of learning and experience. They can learn more because they command more capital to use in their trade as capitalists. With more capital they can attract more highly skilled labor from around the globe. They spend less time and effort on avoiding taxes and interpreting regulations and consulting lawyers and accountants. They can conduct more undetermined experiments, test more falsifiable hypotheses, try more business plans, and generate more productive knowledge. With fewer resources diverted to government bureaucrats, profits are managed by the people who earned them and who thus learned how to invest them successfully.

Seizing resources from entrepreneurs and redistributing it politically is the death of creativity. A tax system based on antipathy to entrepreneurs will necessarily fail to produce the expected revenues because it ignores economic and human reality.

The Technology
Evolution Myth

FLYING TO ISRAEL for a whirlwind visit of technology startups in late 2010, I had stacked on my tray table two recently published books on the sources of technology. Discussions with technologists, their minds in the guts of their machines, would absorb my every waking moment when I landed, but now, seven miles high, I had before me two books presenting an entirely different viewpoint on technology.

The ecstatic jacket blurbs were promising. The first book, *What Technology Wants* by Kevin Kelly,[1] the founding editor of *Wired*, was "the best book on technology I have ever read," said Nicholas Negroponte, himself the author of the crisply definitive *Being Digital*. While Kelly's book looks to the future, its companion, *Where Good Ideas Come From: The Natural History of Innovation* by Steven Johnson,[2] takes a historical approach. Walter Isaacson, Steve Jobs' celebrated biographer, hails Johnson as "the Darwin of technology."

Both Kelly and Johnson offer ambitious Darwinian reflection on machines. Their central proposition is this: human intelligence and its technological artifacts are merely extensions of evolutionary biology, and they are intelligible through the principles of this science. Such an idea might never occur to someone engrossed with the chips-and-fiber, analog-and-digital details, and I well remember the excitement I felt when I first encountered it in George Dyson's incomparable *Darwin Among the Machines*.[3]

Some three decades ago, Dyson was an unknown writer living ninety-five feet above the ground in a tree house in British Columbia. Boasting no academic degrees but powered by a formidable genetic inheritance—his father is the venerable physicist Freeman Dyson, his mother a biographer of Kurt Gödel, and his sister Esther an astronautical software sage—George's mind took flight on the subject of Darwin and technology. He embarked on an inquiry that tends to separate technology from its individual human authors, treating it as an almost independent extension of Darwinian evolution. Eventually clambering down from his arboreal perch, Dyson wrote *Darwin Among the Machines*, a pithy and complete bottom-up theory of technology, which takes its title from an 1863 article by the British polymath Samuel Butler.

Kelly and Johnson are only two of the most notable of Dyson's followers in recent years. The primary goal of these thinkers is to banish any idea of God or a transcendent creator from nature (though Kelly wants to import Him as a panoptic afterthought). A further goal, less explicitly stated but always present, is a devaluation of human creators as well. The theme of their work is "You didn't build that." Individual human beings no more create a technological civilization than God creates the universe. Technology is an independent actor that emerges from the process of biological evolution.

In *Darwin Among the Machines*, Dyson makes the case as well as it can be made. I loved it at the time. But it did not convince me then, and Kelly's new 3-D animated version does not convince me now. When you get down to it, evolutionary biology proves to have almost nothing to say about machines. Separating machines from material evolution is the human mind and consciousness—the capacity to surprise. The attempt to reduce human creativity and technology to Darwinian determinism yields virtually no useful insights.

The argument moves the focus from the human prowess and economic freedom that enable technological progress to the supposedly internal dynamics and tendencies of technology itself. Popular among writers and academics, this approach tends to exalt the Olympian observer of these intriguing processes over the ingenious nerds and entrepreneurs who actually perform them. My own reputation, amusingly enough, has been a beneficiary of this attitude. Because I predicted various developments on the Internet—"worldwide webs of glass and light," the rise of search-and-sort technologies for targeted ads—some have suggested that I was somehow responsible for them. I may have aimed some investment dollars toward those developments, but the prime and crucial movers were the engineers and inventors who actually built the requisite machines.

Unlike many of his followers, Dyson never reduced mind itself to a mere Darwinian artifact. In *Turing's Cathedral: The Origins of the Digital Universe* (2012),[4] Dyson depicts modern computer technology emerging not only from a Darwinian struggle but also from centuries of advance in chemistry, physics, and engineering—and, most crucially, from the overarching mathematical theory of information, at the heart of which is human creativity.

New prophets like Kelly and Johnson, bypassing information theory, construct a false model of invention that ignores inventors and reduces most creativity to details of evolution, "order," and "self-organization." This approach leads to the obligatory worship of coral reefs and rain forests as models of technology evolution, but also to a distorted view of how innovation works.

Johnson's evolutionary perspective allows him to see that inventions have to make use of existing resources and tools or they come "before their time" and fail to gain traction, but it causes him to stumble when he looks at the function of patents. In an ungainly four-sector diagram, he attempts to analyze the benefits of four overlapping environments of innovation: individual versus networked, private versus public. He finds that the most fertile source of new ideas is the fourth quadrant of his diagram—*non-market* (i.e., private) *groups of networked innovators.* "The test," he argues, "is not how the market fares against command economies... but how it fares against the fourth quadrant: the modern research university [where] new ideas"—normally unpatented—"are published with the deliberate goal of allowing other participants to refine and build upon them.... Academics are paid salaries, of course, and successful ideas can lead to much-sought-after tenured professorships, but the economic rewards are minuscule compared to those in the private sector."[5]

Johnson's misguided claim that secretive fourth-quadrant researchers such as Newton and Darwin "did everything in their power to encourage the circulation" of their ideas—the historical record shows that they hid them for decades—raises questions about his view that intellectual property laws are obstacles to openness. The current system of intellectual property law has indeed overreached by including patents on incremental software advances and on "business methods." (The incremental software patents are nearly all obvious to programmers in the same field, and "business methods" are merely the way entrepreneurs compete.) These errors have led to a litigation explosion calculated in a punctilious Boston University study to have reduced the market caps of American technology companies by half a trillion dollars over the last decade.[6] But this folly should not discredit the basic idea of protection for genuine inventions.

The abolition of patents would likely result in a plague of secrecy that retards the process of innovation as classification does in military research. As Edward Teller has observed, democracies are no more effective in conducting secret bureaucratized research than despotisms are. The Soviet Union kept up with the United States in secret nuclear, satellite, and missile technology after World War II. It was the unclassified and commercial computer industry, protected by patents galore, that gave the United States the lead.

Johnson's mistake is assuming that innovation is limited by a scarcity of ideas rather than a scarcity of entrepreneurial capital and the engineering prowess to put them into practice. Mark Zuckerberg deserved control of Facebook, and Brin and Page control of Google, because only they could reify the ideas, making the rather pedestrian concepts work in a treacherous world. The reason capitalism succeeds is not merely that it fosters the creation of ideas but that it can bring those ideas to the marketplace. The system efficiently capitalizes inventions, which are always abundant (hence the throngs of disgruntled inventors) but difficult to sort out and prioritize.

Capitalism assigns to the winners of previous profits the right to reinvest them in new ideas. Profits are crucial because they capture the unexpected upside of successful innovations beyond the predictable yields of the interest rate. It is a knowledge system that preserves the experience of entrepreneurs by allowing them to continue their work and expand their investment if their project succeeds.

Among evolution-of-technology thinkers, Johnson at least attempts to derive practical lessons from his ideas. What is one to make of Kelly's notion that everything from galaxies to starfish to the human mind—and its inventions—is somehow both "contingent" and "inevitable"—that the iPad essentially springs from the laws of physics at the origin of the universe?

Kelly is an able analyst of technology and a gifted poet of science, and some aspects of his book shine. He does a superb job in refuting the "precautionary principle," which he shows to be effective for only one use—stopping technology. "Every good," he says, "produces harm somewhere, so by the strict logic of the precautionary principle no technologies would be permitted." He shows that, in the past, precautionary bans of DDT and asbestos caused more damage than the original threat. Over-reactions to threats are common, he notes—"adding security forces at an airport can increase the number of people with access to critical areas, which is a decrease in security." These are "substitute risks" that "arise directly as a result of efforts to reduce hazards."[7]

"The risks of a particular technology have to be determined by trial and error in real life," Kelly writes—dangers become evident only with scale. If certain technologies pose huge dangers in the hands of enemies, suppressing the technologies only reduces our knowledge of them, thus increasing their danger. It is better to increase our experience with the technology and learn how to neutralize it. Eternal vigilance as the price of liberty produces what Kelly calls a "proactionary principle," emphasizing "provisional assessment and constant correction." Risks are real but endless and have to be prioritized, treating technologies symmetrically with natural hazards such as weather, earthquakes, and meteors.

But when Kelly names his book *What Technology Wants*, he is serious. He concedes, "I don't believe the Technium [his word for all human culture and artifacts] is *conscious* at this point." But he finds deep significance in a robot from a startup called Willow Garage in Silicon Valley that can look for a socket and plug itself in when its battery dies. "It's not conscious, but standing between it and its power outlet, you can clearly feel its want."[8]

Perhaps we should pause a moment with Dyson's discussion of this issue, which Thomas Hobbes and Wilhelm Leibniz addressed in the 1670s, when philosophers struggled to find the "invisible ingredient that moves from the predictability of logic"—for example, the software programs that inform the Willow Grove robot—"to the unpredictability of mind."

Like Ray Kurzweil today, Hobbes thought that the brain was simply a complex machine—that thought was simply a material phenomenon explicable in terms of neural cogs and connections. Leibniz responded by envisaging a brain expanded to the size of a mill that one could walk through. "Now on going into it he would find only pieces working upon one another, but never would he find anything to explain Perception." There would be cogs and connections but no cognition.

In the 1980s, John Searle of the University of California, Berkeley, explained this concept with the image of a large enclosed room filled with clerks following a list of rules in an algorithm for the translation of Chinese words that were submitted to them through a slot.[9] The clerks would seem to translate Chinese, as a computer does, but they would have no more grasp of the language than would the computer.

Kelly follows Kurzweil in speculating that the *quantity* of switches or transistor circuits in a computer can reach a tipping point where it engenders the *quality* of intellect and consciousness.[10] But a computer is essentially a mathematical logic machine, and as Kurt Gödel, Alan Turing, and their followers definitively showed, all such systems are incomplete and depend on outside premises. A maze of on-off switches and branches, no computer can generate its own symbols. Without the presence of real minds, no network of switches, however dense or intricate or combed with feedback loops and homeostatic balancers, can "emerge" as a willful conscious entity. All these devices are symbolic processors, and the symbol systems and codes must originate outside the machine.

The heart of *What Technology Wants* is a huge generalization: "The technium can really only be understood as a type of evolutionary life... [that] can be traced back to the life of an atom... created in the fires of the Big Bang."[11]

As Dyson showed, such ideas, in more modest form, were around when Charles Darwin was working and even before. Beginning with Hobbes's assertion that "collective human association [in the social contract] signals an end to the illusion of technology as human beings exercising control over nature, rather than the other way around," Dyson extends the insight to our own time: "Everything that human beings are doing to make it easier to operate computer networks is at the same time, but for different reasons, making it easier for computer networks to operate human beings."[12]

This inability of modern techno-philosophers to grasp the difference between machines and human beings is a great *trahison des clercs* of our time. They would have us believe that inventors and entrepreneurs are passive figures in an evolutionary process that these self-serving intellectuals alone transcend. They must transcend it, after all, if they can see through it and unveil the great illusion of human creativity.

The root of the failure of the biocentric vision is a misunderstanding of information theory. All the exponents of biocracy, from Dyson and Kelly to Johnson and Kurzweil, tend to mistake information for order. Kelly calls the mind a "highly evolved way of structuring the bits of information that form reality. That's what we mean when we say a mind understands; it generates order."[13] He compares inventions with snowflakes and crystals, which are indeed examples of order. But information and order are not the same thing; they are opposites. Entropy, as we have seen again and again, is not order and predictability; it is disorder and "surprisal," registered as unexpected bits. Bits in a message that are fully anticipated convey no "information"—the receiver already knows the content of the communication.

By blurring information and order and utterly ignoring profit, both Kelly and Johnson fail to recognize the line of demarcation between evolutionary forces and human creations. Evolutionary forces may generate crystals, the stripes on a zebra, vortices, fractals, and other repetitive patterns that can be produced by a simple algorithm (in the case of the zebra stripes, a tiny number of nucleotides). But these phenomena lack complexity measured by informational entropy. As the Princeton economist Albert Hirshman has written, "creativity always comes as a surprise to us."[14] It is a high-entropy event. Innovations are not an expression of equilibrium and order, like crystals or snowflakes, but disruptions of it.

Recall the rule of information theory: it takes a low-entropy carrier (no surprises) to bear a high-entropy message (creative surprises and inventions). In biology, the low-entropy carrier is the regular sugar phosphate backbone of the DNA molecule, which bears the high-entropy codes of life in its nucleotides. Low-entropy carriers with no surprises allow the message to be distinguished from the medium at its destination. Order, which is expressed in law, tradition, predictable politics, stable families, and reliable currencies, is necessary for an innovative society full of unexpected discoveries.

Kelly and Johnson are also mistaken in their belief that innovations chiefly seek efficiency, which is a measure of order and evolutionary incrementalism. The importance of innovations, as Peter Drucker has stressed, is not their efficiency but their effectiveness.[15] They do not do existing jobs better; they redefine the work. They don't do things right; they identify the right things to do. Innovations are not linear but saltatory.

Like Paul Romer, Kelly and Johnson see invention as a process of material recombination, generating new assemblages of atoms or chemical elements in the "adjacent possible." Entrepreneurial creations, they maintain, are mutations, selected naturally by the "market." Entrepreneurs are merely responding to external stimuli in a bottom-up process resembling both Darwinian evolution and Skinnerian stimulus and response.

In real life, however, invention comes first, not the market. The manufacturer of carriages or vacuum tubes or typewriters did not fine-tune components in the adjacent possible. He conceived a new system.

The automobile or transistor or computer absorbs any existing components into an altogether higher-level machine. Indeed, as Drucker has written, to replace an existing system, a new invention must be at least ten times better.

Even the truest believers in the sufficiency of Big Bang biology may question its extension to the development of the Xbox or the cruise missile or even Johnson's and Kelly's books. Biological evolution is simply a forced and unfertile analogy for technological progress, entirely missing its essence, which is human consciousness and creativity.

I contemplated these issues as my El Al 747 approached Israel. To Kevin Kelly on a mountaintop near Silicon Valley, Steven Johnson just north of him in Marin, and George Dyson even farther up the West Coast in a kayak on Bellingham Bay, technology is an option that enhances choice. In Kelly's words, it imparts an "exotropic" bias to the evolution of the "technium." Quoting Brian Arthur's aphorism that "problems are the answers to solutions," Kelly nonetheless affirms that, amid many ambiguous advances, technology ultimately achieves more good than harm.

On this flight to the middle of a war zone, I cannot contemplate technology with such detachment. Technological advance, for beleaguered Israel, is not optional or ambiguous; it is indispensable to survival. It is not a matter of serendipity or "exotropy" or self-organization but the fruit of the mastery of such disciplines as mathematics, chemistry, algorithms, metallurgy, condensed matter physics, molecular biology, optics, digital signal processing, medicine, engineering, and—indispensably—information theory. Perhaps all these subjects originate in the Big Bang, but nothing in evolutionary theory illuminates the connections. For the same reason that chemistry cannot be reduced to physics—the density of information is much higher—biology cannot be reduced to chemistry, or human creativity to biology.

Here we encounter what the atheist and philosopher Daniel Dennett calls the "universal solvent" of natural selection. With its reductionist acids it breaks down all larger systems into components and elements and blends, removing purpose, structure, teleology, and creativity. Kelly tries to retrieve it all at the end of his book with rhapsodic rhetoric that lends an almost divine nimbus to his "technium." But the previous 320 pages of *What Technology Wants* are devoted to eliminating all hierarchy in the universe through a bottom-up, material model that devours even the human mind in its Darwinian maw. As the Nobel laureate Max Delbrück once commented, attempts to reduce the brain to material flux reminds him of nothing so much as "Baron Munchausen's effort to extract himself from a swamp by pulling ever harder on his own hair."[16] Another Nobel laureate, the physicist Robert Laughlin, noted the futility of this project: "Much of present day biological thinking is ideological"—that is, the explanation "has no implications and cannot be tested." Such ideas "*stop* thinking rather than stimulate it." The answer is always "Evolution did it."[17]

Hard technologies, meanwhile, are in eclipse. Johnson's idea of a breakthrough is Twitter. Venture capital's focus on the trivialities of social networking is an ominous portent for the future of American world leadership and defense. The global warming hypothesis might make Johnson's list of the greatest ideas of the last two hundred years and cause Kelly to abandon his critique of the precautionary principle. But these fears are not science, and no other political campaign has been so destructive to the innovation process.

The heroic drama of innovation—its rigor, genius, and discipline—goes on where economic and political freedoms permit it and military and industrial exigencies promote and protect it. But fashionable thought in the United States—represented by Kevin Kelly's and Steven Johnson's books—encourages our technological decadence.

Israel: InfoNation

IN EARLY 2012, Apple Computer purchased a tiny, five-year-old Israeli company for between three and four hundred million dollars. Headquartered in Herzlia, Anobit made "flash" memory systems, consisting of memory chips that keep their contents when the power goes off. "Nonvolatile memories," as they're called, are vital to Apple's miniaturized wireless devices—iPod, iPhone, iPad, and MacBook Air—whose huge data storage capacity, comparable to a desktop "hard drive," must reside not on discs but on chips.

A forte of Israelis since Dov Frohman created the first non-volatile memories for Intel in 1971 and then brought the technology back to Israel in Intel's Haifa design center, these tiny chips are arrayed in "solid state drives." They are quieter, faster to access, and more durable and shock-resistant than mechanical hard drives.

Anobit ("another bit") doubles the capacity of a flash memory system and extends its lifetime, minimizing the tradeoffs between cost, capacity, and endurance. Anobit can make the memory systems in Apple products half the size, or dramatically cheaper, enhancing the device's portability, longevity, and even elegance.

Based on information theory, Anobit's technology illustrates one of the key characteristics of innovation in the information age. Anobit makes no physical changes to the memory chips. It supplies an algorithm on the memory controller that enables more efficient error correction and adaptation, optimizing a chip's performance. This algorithm, which has many other uses as well, makes it possible to double the number of bits in a memory or across a motherboard—and potentially even in an Internet router or switch. Anobit is an example of the power of information theory, and it is also a vehicle of the new economy based on Shannon's ideas.

A few weeks after Apple's acquisition of Anobit, AT&T announced a solution to the dire capacity problems afflicting its wireless network, where traffic had been doubling every year and was on track to rise sixfold. After the iPhone launch in 2007, a flood of data-intense devices drove traffic on the AT&T network up an estimated twentyfold over several years. An iPhone uses twenty-four times the bandwith of an ordinary cell phone, and an iPad uses 122 times that bandwidth. AT&T customers were suffering dropped calls, intense network congestion, and acute bandwidth bottlenecks. Analysts were speculating that AT&T would lose its Apple contract.

At first, the telecom giant pursued a conventional solution, a physical solution. The company would have to purchase more spectrum, which is increasingly scarce and expensive, or deploy thousands more cellular base stations, new amplifiers and antennae, and complex systems of tiny base-station extenders called picocells. The cost to renew the entire network would be scores of billions of dollars.

Fortunately for AT&T, we live in an age when information theory overcomes physics. They turned to the Israeli company Intucell, which could accomplish the same goals at a cost of fifty *million* dollars rather than fifty *billion*. Like Anobit, Intucell employed information theory to solve problems that previously required dramatic expansions of physical plant and equipment.

Intucell's software turns the entire network into a single adaptive system—a "Self-Organizing Network." Constantly gauging network conditions, it adjusts transmission power in every cell to optimize both its coverage and capacity. As the telecom writer Kevin Fitchard describes it, "Quite literally cell towers start following you, expanding their cell radiuses as you move closer to their edges, while neighboring cells recede. By moving the network around you as you yourself move through the network, [Intucell's self-organizing network] can find the optimal overall topology at any given moment to provide the best coverage and capacity to thousands of users within a cluster of cells."[1] Virtual networks also improve management. When a single cell site fails, a frequent occurrence at AT&T, the Intucell system can expand surrounding cells to pick up the traffic. When the problem is resolved, the entire system can revert seamlessly to the original configuration.

Having tested the system since April 2011, AT&T was approaching full deployment by the end of 2012. The result has been a 10-percent decline in dropped calls, as much as a 40-percent increase in capacity, and as much as a megabit-per-second increase in bandwidth for the average customer. Early in 2013, Cisco bought Intucell for some $400 million.

Wireless telephony has been the spearhead of the American and world economies for the last decade and is a likely source of huge growth in the next. Smartphone sales rose from 300 million units in 2011 to 712 million in 2012 and are projected to reach two billion over the next four years. These are genuine Turing machines, ready to do anything. An explosion of wireless data is predicted—from ninety petabytes per month in 2009 to six thousand petabytes per month in 2015—up seventy-fold in seven years. Apart from Qualcomm in San Diego, Israel has been leading the world in this field, exploiting its edge in information theory algorithms nimbly launched by start-up companies.

For decades, the holy grail of wireless telephony has been software-defined radio, which would substitute information for physical advances in microchip geometries. Working at a company called ASOCS (Adaptable Silicon on a Chip), a team led by Simon Litsyn, an immigrant from Russia, and his student Doron Solomon have developed a virtual communications platform that can perform all the needed processing functions for a wireless system in instantly adaptable software. One of their first products is an adapter card for China Mobile that enables a laptop concurrently to stream a TV program from a satellite, to send a text over a GSM link, and to conduct a Skype call over the Chinese broadband standard TDS-CDMA. A technology with important military uses, the ASOCS architecture replaces an array of chips or hardware processing threads. It is the epitome of an information theory scheme that replaces costly physical-layer complexities.

Still a small start-up, ASOCS is just one of many Israeli companies pioneering information systems that outperform complex hardware. Provigent (a recent acquisition of America's Broadcom) has developed unique algorithms to eliminate a serious bottleneck between cell phone base stations and the network. AMIMON controls lossless wideband video technology that eliminates wires from some fourteen thousand surgical rooms in American hospitals (with cooperation from the U.S. medical manufacturer Stryker).

These are only a few of the thousands of Israeli companies operating on the leading edge of the information economy. Involved in everything from biotech to cloud computing, these companies are proof of the continuing importance of information theory and practical embodiments of entropy economics.

The Israeli advantage is partly demographic. There are perhaps a million immigrants from Russia who, previously barred as Jews from manufacturing jobs in the Soviet Union, gravitated to higher mathematics and software. Taking advantage of their mathematical and algorithmic skills, Israel is consolidating its global supremacy, behind only the United States, in an array of leading-edge technologies. Beyond wireless telephony and computation, it is the international master of microchip design, network algorithms, and medical instruments.

The effects of innovation in information processing extend far beyond computer and communications companies. In a world beset by water crises, Israel is incontestably the leader in water recycling and desalinization. Since the state was founded in 1948, its population has grown tenfold, its arable land threefold, its agricultural output sixteen-fold, and its industrial output fiftyfold, yet its net water usage has *dropped* an astonishing 10 percent. This unique achievement is the result not of sanctimonious laws or disruptive environmental litigation but of combining information with enterprise. Israel gets half its water from desalinization, recycles 95 percent of its sewage, and employs extraordinarily efficient high-tech irrigation systems. It pumps brackish water from deep beneath the desert in Arava to sustain an agricultural miracle. All these technologies benefit from sophisticated software that responds intelligently to the flow of information, sensing and repairing leaks and targeting drip irrigation nozzles directly at the roots of plants and combining them with nutrients. Neighboring Egypt and Iran, meanwhile, are the world's highest per capita water users.

Information tools are becoming the most effective form of military power in an epoch when the world's great cities from Seoul to New York face a threat of terrorist rockets. Israel's answer to that threat, the newly battle-tested "Iron Dome" and Arrow anti-rocket systems, are based on software and microchip innovations that radically reduce the weight and cost of interceptors. Its information-intensive defenses are cheaper and more effective than schemes that throw money and hardware at the problem.[2] Missile defense is cost-effective because of systemic advances that allow the equipment to avoid jamming, differentiate payloads from chaff in real time, and avoid wasting interceptors on misdirected incoming rockets.

Much of the world believes that Arab Palestinians have a compelling claim to major stretches of Israeli land. But the land of Palestine could not have supported one-tenth the population it does today without the heroic work of reclamation and agricultural development by Jewish settlers beginning in the 1880s, when Arabs in Palestine numbered a few hundred thousand.

At the heart of the Israeli accomplishment is mastery of information tools—replacing land, labor, and capital with the power of knowledge.

The Knowledge
Horizon

IN THE WAKE of the financial crisis, a number of books have appeared suggesting that America's economic woes are the result of the decline of innovation, which stems from the exhaustion or corruption of science. Tyler Cowan's acclaimed broadside, *The Great Stagnation*, written with the collaboration of the investor Peter Thiel, contends that the current fabric of innovation is faulty. Phony innovation "often takes the form of expanding economic and political privilege, extracting resources from government by lobbying, seeking sometimes extreme protections of intellectual property laws, and producing goods that are exclusive or status related rather than universal, private rather than public...."[1]

Summarizing his case, Cowan writes, "Think of twenty five seasons of new fall season Gucci handbags." This idiom is familiar to observers of economic fashion over the last fifty years. John Kenneth Galbraith was the master of elegant Gucciphobia, beginning in *The Affluent Society* with a similar distinction between public and private goods and a similar charge that the United States was stinting on its public environment in an orgy of private consumption. But Cowan, observing our vast government bureaucracies, believes that public spending is even more otiose and self-indulgent than private spending.

A former Bell Labs engineer and venture investor, Andy Kessler, sold out at the top in 1999 to become an influential pundit and best-selling author. He has a different perspective on the private side: "Just about every product or service that makes our lives better requires a mass market or it's not economic to bother offering...."[2] This is a lapidary observation. But Cowan and Thiel take their argument another step. It is not just that the rich have captured innovation, a false charge for the most part, but that it is dying in captivity to government.

Cowan cites a study by a physicist at the Department of Defense, Jonathan Huebner, that charts innovation events per capita from the end of the Middle Ages. Innovation peaked, by Huebner's measurement, in 1873.[3] Since then the rate seems to have tumbled from twenty events per year per billion persons to a mere seven in the late 1990s. Huebner also shows that in proportion to national income and educational outlays, innovation is more sluggish than in the nineteenth century. We lavished money on research and development between 1965 and 1989, doubling spending in the United States, tripling it in West Germany and France, and quadrupling it in Japan. But in all these countries, growth slumped and patents reached a plateau. The United States, for example, according to one measure cited by Cowan, had more patents in 1966 (54,000) than in 1993 (53,000). Although these numbers are hard to credit in view of far higher patent bureau reports, it is true that per "researcher," patents have plummeted.[4]

I lived through this era myself, arriving in Silicon Valley in 1980 to write a newsletter on semiconductors for the venture investor Ben Rosen and the venture thinker Esther Dyson. Cowan's figures don't ring true to me. I found myself in the midst of a cauldron of creativity in microchips, software, computers, and semiconductor capital equipment. While the number of patents granted annually was soaring toward 100,000 in the 1980s and 200,000 in the mid-1990s, most of the technological advances remained unpublished and unpatented. It was concealed as trade secrets and tacit knowledge, what I call *latents* (from the Latin for "hidden") to distinguish them from patents (from the Latin for "open").

Then I was engulfed by fiber optics. I proceeded to chronicle the booms in photonic and wireless technology in the 1990s for *Forbes ASAP* with Rich Kaarlgard. As a result of my writings, I am depicted, in the pages of Wikipedia, as a "techno-utopian futurist." But a utopia, by definition, does not exist. We live today in a cornucopian, techno-topian world.

On the surface, though, the Thiel-Cowan Thermidor must be true. All economic growth ultimately stems from innovations. When growth flags over extended periods, innovation must necessarily be slowing. Even with our sluggish economy, though, Huebner's chart remains unconvincing. It is dominated by population growth in the denominator rather than the tiny selection of innovation events that he presents as a numerator. Huebner's numbers do not say innovation has declined; indeed, the number of innovation events is up eleven-fold or more. He just shows that it has slowed relative to the population, to education spending, and to spending on research and development.

Economic growth and technological progress are an elite phenomenon, driven by a tiny minority. Educational spending has little to do with it. In my 2009 book *The Israel Test*, I chronicled the entrepreneurial achievements of one small group, approximately twelve million Jews, over the last century. As Charles Murray documents in *Human Accomplishment*, a minority of this minority of Jews has outperformed every other identifiable group in scientific and entrepreneurial accomplishment around the world for most of the last hundred years.[5] Explosions of economic advance attended their every movement across the landscape of the twentieth century, from Vienna and Budapest in Central Europe to New York and Los Angeles in the United States to Tel Aviv in the Middle East.

Ludwig von Mises noted in 1940 that in his native Austria, there were about a thousand men who were capable of organizing production for export. "At least two-thirds of these one thousand men were Jews.... They are gone, scattered around the world, and trying to start again from scratch."⁶ Also left having to start from scratch was the Austrian economy. It never fully recovered. A similar story can be told of Budapest, a city led by a small cohort of entrepreneurial Jews to commercial preeminence in the late nineteenth and early twentieth centuries. The banishment and expropriation of Jewish entrepreneurs brought Budapest's industrial miracle to a halt. The flight of Jews from Germany and from the Soviet Union resulted in the rapid deterioration of these nations' innovative genius. The arrival of millions of these German and Russian and Polish Jews in the United States made possible the Manhattan Project that ended World War II. They populated Israel, reanimated its economy, and spurred an intellectual-led boom in the cities where they settled. Without the diaspora, the information economy of Silicon Valley and San Diego, the entertainment industry of Los Angeles, and the finance and fashion centers of New York would lie fallow and infertile.

Much of the misunderstanding of economic growth stems from incomprehension of its intrinsic elitism. Huebner's chart registers a massive surge of population—in recent years chiefly in China, India, and Latin America—which corresponds to the health-care and hygiene revolutions achieved mainly by American innovators. Supply of health technology creates its own demand. Just as the rise of entrepreneurial creativity drove the division of labor, so the rise of technology fuels population growth. Both phenomena reflect the contributions of a tiny minority of the earth's population.

The diagnosis of an innovation slump is popular because it provides an innovation alibi—the insistence that inventions are an autonomous force, outside the economy, an externality born of science and technology and largely beyond the reach of policy. All the government officials filling the channels of enterprise with noise can claim that the dearth of innovations is not their fault. The scientists, engineers, and entrepreneurs are failing, and it's time for another round of redistribution. Nearly every prominent analyst of capitalism has fallen for the myth of the opportunity dearth. Throughout the history of economics, there has always been someone predicting that the flame of innovation would die and the system would settle into a "stationary state."

Thirty years ago, I wrote in *Wealth and Poverty,*

> Sociology does not recapitulate biology. Even though senescence may afflict great men on their paths of glory to final equilibrium, it is not a characteristic of nations and capitalism, and capitalism, like the family, is not an institution that can become obsolete or decrepit as long as human societies persist. Human needs and numbers annually increase; science and technology provide their continuing surprises; the exigencies, complexities and multiplicity of life on earth become yearly more unfathomable to any tyrant or planner. No nation can grow and adapt to change except to the extent that it is capitalistic, except to the extent, in other words, that its productive wealth is diversely controlled and can be freely risked in new causes, flexibly applied to new purposes, steadily transformed into new shapes and systems. Time itself means continuous change of knowledge and conditions. Among all states, it is the stationary state so favored by the prophets that is most sure of withering away.[7]

Innovation is always a product of individual innovators, a rare and dynamic breed not always appealing to the millions who depend on their creativity for their own comfort, health, and security. One of the peddlers of the innovation alibi, though he may deny it, is a leading innovator himself: Peter Thiel, the billionaire titan of PayPal, Facebook, Palantir, and other companies. Ensconced in a neo-classical financial temple in the Presidio near the Golden Gate Bridge, he is the venture capitalist supreme, the man who led what he seems to believe was a final Silicon Valley boom. Early in the new millennium, he announced the exhaustion of the creative founts of Western ascendancy. He predicted the slide into economic decadence that must follow a dearth of invention and the corruption of science by politics.

Apart from a nanoparticle sheen on fabrics that can shed spilled milk and honey and a providential supply of Viagra for a nation whose hotel industry makes most of its profits from hardcore pornography in every room, where are the major new technologies? Where are transformative breakthroughs comparable to the space flights of yesteryear, the transcontinental railroad, the internal combustion engine, the nuclear power plants, the antibiotics, the radios and televisions, the three-dimensional films, the silicon microchips—all radical novelties that reshaped the lives of our parents and grandparents?

To Thiel, today's advances seem puny by comparison. Can anyone suggest with a straight face that Facebook and PayPal are equivalent to these earlier wonders? Web 3.0? Or even their formidable precursors, Google and Amazon? Neal Stephenson, the novelist and techno-seer who has best captured in literature the spirit of the age, answers "no." In a dour essay, "The End of Big Ideas," he asks why we can no longer scale the colossal peaks of yore—the moon landings, the Cold War victories, the genome mappings, the worldwide webs of fiber optic light, the titanic towers looming over shining cities?[8]

Are the forces of innovation exogenous and providential? Or are they the results of human genius and creativity, enabled by societies that affirm them? Is there really an innovation alibi for the forces reducing man to an economic homunculus, settling into a stagnant world?

Thiel issues a challenge: "Pharmaceuticals, Robotics, Artificial Intelligence, Nanotech—all these are areas where the progress has been a lot more limited than people think. And the question is why?"[9] To answer Thiel, let us return to Richard Feynman's legendary lecture to a Caltech seminar back in 1959. Titled "There's Plenty of Room at the Bottom,"[10] it entranced generations of engineers and technologists with a vision of radical advances in the miniaturization of all technology. It began a mode of thinking small. Envisioning atomic-scaled computers and infinitesimal memory devices, inscribing the *Encyclopaedia Britannica* on the head of a pin, Feynman's speech inspired Silicon Valley. At the time, the industry was still laboring with discrete transistors combined on foot-long transistor-transistor logic boards.

Feynman's speech was galvanizing, visionary, and influential. It prophesied much of the microchip industry. But Feynman's chief concern was not the prediction of microprocessors and flash memory chips. It was the special ambition that we call "nanotech"—the erector-set vision of nanofabrication. On that point Feynman was wrong, both in the practice and in the science. Building machines at the nano-level, atom by atom, in the trillions, amid gigahertz vibrations of Brownian motion and gigantic viscosities and stickiness, turned out to be a snare and a delusion.

Like many in the world of twentieth-century physical science, Feynman was captivated by the materialist superstition—the idea that the world can be understood from the bottom up by breaking it down analytically into ever smaller pieces. A veteran of the Manhattan Project and a recipient of the Nobel Prize for his work in quantum electrodynamics, Feynman adopted a position of supreme physical reductionism. He envisaged tweezers moving atoms or even sub-atomic particles one by one to create new chemical elements from the bottom up. He imagined systems of nanotech manufacturing that bypassed biology and chemistry to build new machines and materials. He even speculated about nano-replicators—nano-devices that could reproduce themselves. None of these goals is close to being achieved.

Frustrated in its aspirations to the godlike composition of molecules, physics has reached a dead end, an aporia. Its confession of intellectual exhaustion is a theory called the "infinite multiverse"—an infinite number of parallel universes. Disguising an inability to account for our one fabulously complex cosmos, the multiverse theory is an illegitimate extension of Feynman's work mapping the paths of electrons by assuming that they take all possible routes. In a brilliantly successful stratagem, he calculated the actual path by allowing all the interference effects to cancel, leaving the residual as the solution. Multiverse theorists assume that all the electron paths actually exist in parallel universes. As time passes, popularizers such as Brian Greene and Michio Kaku and formidable physicists such as Steven Weinberg and David Deutsch assert with ever more confident gullibility the case for this canard.[11] They have reached the end of the line in actual physical explanations of nature, so they explain it away.

Infinite multiple universes are a prime example of what Whitehead called the fallacy of "misplaced concreteness," ascribing a physical reality to an abstraction or a model. But as Richard Dawkins puts it, "Better

many worlds than one God." To some adherents to the materialist superstition, the multiverse has the virtue of explaining human life and consciousness through an anthropic principle. Among all the universes, we
can observe only the one that accidentally produced us.

The philosophical stultification of physics was repeated in biology,
where Darwinian materialism has reduced pharmacology to accidental
processes. Robotics has actually been a major triumph, yielding promising developments like Google's fleet of automated cars, "lights out"
factories, and robotic surgery. But the field has been shadowed by unful-
filled expectations of a venture capital breakthrough or the creation of
humanoids.

The information theorist and mathematician Gregory Chaitin of Bell
Labs explains the frustrations of nanotech, pharmacology, and artificial
intelligence through his calculations of the information content of known
physical, chemical, and biological laws. By measuring the entropy and
the complexity of the shortest computer programs that can generate these
laws, he measures the relative information content of scientific disciplines.
As Chaitin demonstrates, physics cannot explain chemistry, because the
information content of the currently known laws of physics is dramatically smaller than the information content of chemistry or biology.[12] The
latter fields too suffer from a materialist superstition that leads their
practitioners to overestimate their promise.

Reaching a philosophical dead end, these sciences have largely failed
as sources of new technology. Richard Feynman, though, pointed the way
to a solution. At the end of his life, he turned away from an increasingly
filigreed physics to join his son at a company called Thinking Machines
in Cambridge, Massachusetts, pursuing the still-vibrant science of information theory, which continues to thrive in new and more fruitful forms
that both explain and transcend the limits of materialist science.

While modern physics has become sterile, information theory produces an ever-growing tide of new technologies and venture capital successes. Feynman's late-in-life move to Thinking Machines pointed the
way toward the scientific source of twenty-first-century technological
advance in general, from robotics and artificial intelligence to pharmaceuticals and even nanotech. Rather than pursuing the new physics, old

chemistry, and homiletic biology to their materialist dead ends, innovators increasingly ascend to a higher level of abstraction unifying all sciences on a new frontier of information theory.

Thinking Machines produced the Connection Machine, a massively parallel supercomputer that did not succeed commercially but that epitomized the information sciences animating current technological progress. Its founder, Daniel Hillis, and its computer architect, Guy Steele (now at Apple), were at the heart of the creative ferment in computer science at MIT and Silicon Valley. Feynman felt at home in their company.

For the last two decades, world industry has been moving "up the stack" of computer technology. The "stack" is a seven-layer model of computing and communications. In previous decades, most of the action occurred in the bottom layer, which is the physical layer. It comprises all the microchips and printed circuit boards, racks and wires, fiber optic lines and lasers, amplifiers and sensors, iron oxide discs and "gorilla glass" that make up the material infrastructure of computer and network systems. In other words, the physical layer represents everything you can see—all the physics and chemistry of the machine.

But what you see is not what you get. Above the physical layer are six levels of invisible but vital abstraction—layers for everything from "data links" and transport to "presentation" and applications—that bring the physical layer to life and account for the bulk of the commerce and technological progress of the information age. Without the higher layers, the computer is just an expensive and fragile brick.

This pattern of layered hierarchies is found in biology as well. In both domains, the physical layer relies on the layers higher in the stack when it performs any of its functions. These higher layers are independent of the physical layer and can be coded and stored on a variety of physical media, from paper to polymers to whirling discs and optical memories. DNA is the physical layer for storage of the intricate patterns of nucleotide "bases" and other physical systems that encode the genetic information embodied in a human being. This genetic information can be converted into binary computer language and stored in a computer memory, just as the contents of a computer memory could be translated into language that might be memorized by a sufficiently diligent and motivated human being.

In 1958, Francis Crick, the co-discoverer of DNA, formulated what he called the "central dogma" of molecular biology. (He later acknowledged that he hadn't understood the meaning of "dogma" and that it was a poor choice of word). It describes the flow of genetic information from DNA to RNA to protein, a sequence that cannot be reversed. The central dogma, that is, bars the genetic code of a particular creature from being reprogrammed by its physical layer, its proteins. The "word" shapes the flesh, but the flesh cannot change the genetic word. DNA codes can define amino acids that form proteins, but proteins cannot specify the codes that define them. This is a rule not of physics or chemistry (changes are physically possible) but of the information in the relevant codes.

The central dogma also applies in computer science, defining the hierarchical rules and boundaries of the stack. A researcher could know the location of every molecule in a computer, but without knowledge of its software codes, he would know little or nothing about what the computer was doing. The directional rule of the central dogma stems from the structure of thought. In the mid-1930s, Alan Turing conceived in his mind an abstract generic model of a universal digital computer, on which all other digital machines, whether made of silicon, gallium arsenide, or Lego blocks, are conceptually based. But no combination of silicon and exotic chemicals in the physical layer could create an abstract Turing machine, which is above any of its particular material expressions in the hierarchy.

Advances in the engineering of physical-layer technology continue to extend the reach of Moore's Law. New photolithography equipment can write on a microchip as small as ten nanometers, and complementary devices implant ions and dope surfaces with active elements. Such achievements constitute the real nanotech. But enabling all this first-layer equipment are many layers of ingenious software. The slowing returns from Moore's Law engineering will not produce the next generation of blockbuster high-tech ventures. Those businesses will be the result of "up-stack" inventions—new software and systems based on the science of surprise, as information theory takes on ever more sophisticated and effective forms.

The pharmaceutical industry repeated the error of nanotech—materialism's superstitious faith in Darwinian randomness, which blinds it to the huge promise of information theory. Big pharma has adopted a trial-and-error method of new drug discovery that fails to accumulate knowledge in learning curves, and focuses its systems on chemistry and physics rather than on genetic information.

The prevailing technique for searching for promising molecules is serendipity, hardly more systematic than the accident that uncovered penicillin by the observation that molds on a loaf of bread could kill bacteria. Today drug researchers throw a million candidate molecules at the DNA sensor grid of an Affymetrix "GeneChip," an amazing device for reading the DNA of biological materials. Nonetheless the vast complexity of genetics dooms some 99.99 percent of the resulting data to uselessness.

Ignoring the central dogma, the learning curve, and the entropy standard of information, big pharma proceeds from one ad hoc chemical combination or molecular experiment to another. Testing millions of molecules, it sorts out the ones with a possible therapeutic effect. This process might seem to be awesomely productive, until you recognize that the number of possible molecules outnumbers the atoms in the universe. The search remains a million tiny shots in a dark sky.

Employing the information theory of biology, an Israeli company named Compugen is changing this process from accidental observation with no learning curves to high-throughput, planned observation with feedback loops and learning. In conformity with the central dogma, it has developed a top-down model of the DNA information cascade from genome to proteome to the small-molecule "peptidome" of pharmaceutical candidate elements.

Compugen's researchers have found that the genetic cascade is far more complex than was previously believed. It is fraught with unexpected feedback loops and redundancies. This chain of information incorporates much of the so-called "junk DNA" that was dismissed as detritus of random processes by previous researchers with an accidental view of evolution. Based on this proprietary science, the Compugen system permits a purposeful search for relevant high-entropy peptides from a vast range of possible choices. It builds models and algorithms based on predictive biology—up the stack of manipulable information, rather than in the unforgiving processes of chemistry and biology at the physical layer.

The proof of the Compugen model of the DNA cascade from genes to transcripts to amino acid to useable peptides is Evogene, a wildly successful agricultural genetics spinoff from Compugen. Evogene has designed hundreds of plants with specified characteristics—protection from insects, endurance of drought, resistance to bruises, desired shapes and sizes. In early 2012, Evogene signed a deal with Monsanto worth close to $50 million under which the American agricultural science leader will move most of its future research and development to Evogene in Israel.

As in pharma, so in the other most fertile fields of innovation. Information theory operates at a higher level of generality and abstraction than do the physical sciences. It comports with the unique powers of human beings and the supreme machines of the age. It will transform every science and technology. People who despair of the future of innovation because of the recent frustrations and fantasies in physics and biology will be startled by the achievements of the coming era.

So where are we? Why all the dismal projections? Frustrated at airports, Tyler Cowan wishes he could be teleported to Kansas or Korea or the moon to live in a Gucci-free egalitopia. Peter Thiel wants supersonic flight and real genetic medicine, robotic vehicles, and new libertarian city-states at sea. Ray Kurzweil pushes for a prosthetic life, an upgraded bionic body with veins vamped with nanobots, chasing down viruses and cancers, repairing outworn tissue and extended by virtual worlds of glass and light. John Holdren (despite being a form of carbon-based life) aspires to electrify the crowds with a minuscule carbon footprint, an environment of national parks and wilderness wetlands, and a chosen people bearing only a few very docile children tele-transported by bicycles.

Neal Stephenson seeks a world worthy of a swashbuckling retro-futurist scrivener like himself, ravishing utopias rather than blue screens or smoky dystopias with dirty snow. Mahmoud Ahmadinajad prays for Islamotopia, redeemed by the twelfth imam rising up from his centuries of dank slumber at the bottom of a well in Natanz, brandishing nuclear weapons and set to unleash a Caliphate in California, free of Jews, Americans, and Sunnis and thronged by furtive women in saggy black shrouds.

To each his own. But under capitalism you do not get what you want. You get what the world's entrepreneurs can deliver in response to the vectors of new technology and desires of the public. All the futurist visions bear an implied critique of our current information age. Some of us have had a bad day. We have botched some of our favorite investments. Some of our ambitions have foundered. But the idea of a technology doldrums is ultimately unfounded.

Cowan wants to teleport. Despite quantum teleportation, physics cannot do it for him and never will. Except in the pages of Michael Crichton, he will not commute among imaginary universes. But information systems can supply him with all the virtual worlds he could want. He can already conduct video-teleconferencing sessions around the globe for free. He will soon be able to conduct these conversations in high definition and three dimensions for a reasonable price.

Thiel complains about air travel, but modern planes—designed and built with the aid of computers and stuffed with computer systems—carry more people than ever before, more safely than ever before, at inflation-adjusted prices lower than ever before. I read on a billboard in Bangkok that half a million people are in the air at this moment. They are also on the air. Invisibly tied to the world through the Internet, they work or entertain themselves while they travel. They land at cornucopian airports that link them swiftly and easily to ground transportation. Yes, they have to make their way through infuriating security systems. But as a result of increasingly sophisticated information tools, those security systems are slowly improving. If we are lucky, and provident, they will succeed in saving a vulnerable infrastructure from its dedicated enemies.

Cowan grouses that the Internet favors a cognitive elite. That is exactly why it is so valuable. The human race triumphs through its cognitive elites. He is probably right in thinking that a new Hitler could not get very far, but if we are to avoid the futuristic visions of Ahmadinedjad, our cognitive elites had better stay on board for the defense of the country and its allies. And we should understand that an indispensable portion of our cognitive elite resides in Israel.

The world is moving from the scientistic guilds and sects of yore toward the new sciences of information. Fragmented and futilitarian, the academic sciences are turning to politics, panics, and cartels to preserve their old privileges. Decades ago I pored through the Harvard catalogue and concluded that 80 percent of the courses stultified their students. Now those stultified students are running the country. Most of the courses they took were either self-evident or wrong, ideological or tautological, twisted or trivial.

Meanwhile, rising to an ever more comprehensive synthesis is the theory of information. Science, nevertheless, is making it increasingly clear that the information is more fundamental than the matter. In 1955, the legendary quantum physicist John Archibald Wheeler formulated the doctrine "it from bit" (thing from thought). "All physics ultimately stems from the information bit." In 2005, a physicist from Massachusetts named Seth Lloyd, who pioneers quantum computers, propounded an elusive but coherent theory that renders a bit of information the very source of matter at the beginning of time.[13] In the beginning was the word … and the word was entropy, an enigmatic concept that links information with thermodynamics, heat with uncertainty, chaos and noise with order and structure, and laws of physics with freedom of will and choice.

During the first half of the twentieth century, the quantum theorists thus cast the definition of matter into the arena of information theory. After World War II, the physicists trudged in themselves—leaving matter behind to be pondered and palpated, titrated and measured with a spectrograph by chemists and geologists. Solidity itself became a manifestation of uncertainty or of fields of force spread out beyond their particular locations with scant materiality.

Now an abstruse and occasionally revealing discipline has emerged called quantum information theory. It is preoccupied with quantum entanglement. Disparaged by Einstein as "spooky action at a distance," it links quantum entities across the universe in instantly propagated information skeins apparently beyond time and space. Its leading practitioner, Christopher Fuchs, presents quantum information theory as "a grand adventure in front of the physics community [which] may well be a key to great progress in fundamental physics."[14] But progress may be slow, he cautions. "In the meantime there is much solid work to be done exploring entanglement, information vs. disturbance, and the information carrying capacities of quantum mechanical systems."[15] Fuchs declares that quantum information theory could not have happened without the pioneering insights of Claude Shannon.[16]

Information theory comprehends and subsumes game theory, computer theory, cybernetics, genetics, network theory, demography, and strategy. It reaches into an array of engineering fields from software programming to chip design, computer architecture, robotics, computer-aided engineering, graphics, and even artificial intelligence. From its roots in Gödel, Turing, Neumann, Morgenstern, Shannon, Post, and Chaitin, it knows its limitations—its inexorable incompleteness, its dependence upon mental processes that it cannot subdue. From its constant tests in engineering, it knows its authenticity and authority. Governed by entropy (measuring freedom of choice), it is a science of human liberty, knowledge and power. And it is generating a golden age of unexpected new technology.

The Power
of Giving

CAPITALISM BEGINS WITH giving. Free markets and exchanges are characteristic of capitalism, but they are a result of entrepreneurship—not a cause of it. It is not the exchange that elicits the goods and generates the increase in their value; it is the initial gift that evokes the desire to reciprocate, and which thus induces exchange. The anthropological evidence, detailed in the original *Wealth and Poverty*, suggests that capitalism begins with the gift and continues with competitions in giving.

A gift will elicit a greater response only if it is based on an understanding of the recipient's needs. As any baffled beneficiary of a costly but unwanted Christmas present can attest, giving is difficult and requires close attention to the lives and longings, tastes and talents of others. In the most successful and catalytic gifts, the giver fulfills an unknown, unexpressed, or even unconscious desire in a surprising way. A successful gift startles and gratifies the recipient with the unexpected sympathy of the giver. In order to repay him, however, the receiver must come to understand the giver. The contest of gifts leads to an expansion of human sympathies.

The circle of giving (the profits of the economy) will grow as long as the gifts are consistently valued more by the receivers than by the givers. In deciding what new goods to assemble or create, therefore, the givers or investors must be willing to focus on others' needs more than on their own. The difference between the value of an item to the giver and its value to the recipient is the profit. Profit is thus an index of the altruism of an investment.

Say's Law defines this sequence of providing first and getting later. In pre-monetary economies, such gifts were a way of escaping the constraints of barter. Without money, all exchanges must be partly predetermined. To that degree, they are devoid of surprise and thus of profit.

In a monetary economy, the capitalist gifts are called investments. Capitalists relinquish resources to others in the hope of receiving surprising transformations, new goods and services, and new value to be reinvested. The essence of giving is not the absence of all expectation of return, but the lack of a predetermined return. Like gifts, capitalist investments are made without a predetermined return. The return partakes of surprisal.

The unending contributions of entrepreneurs—forgoing consumption, exploring the marketplace, investing capital, creating products, building businesses, inventing jobs, accumulating inventories—all long before any return is received, all without any assurance of success, all in response to an imaginative sense of the needs of others, constitute a pattern of giving that dwarfs in extent and essential generosity any socialist scheme of redistribution. In a capitalist economy, every worker and businessman knows in the marrow of his bones that his buying power consists of his supplying power. He values his money because his expenditure of funds is psychologically rooted in his earlier expenditure of effort.

It's not the desire to consume or indulge that chiefly motivates capitalists, but the freedom and power to consummate their entrepreneurial ideas. They are inventors and explorers, boosters and problem solvers; they conduct experiments and accept the outcomes without recourse.

Are they greedier than doctors or writers or professors of sociology or assistant secretaries of energy or chairpersons of the Sierra Club or the Natural Resources Defense Council? Their goals may seem more mercenary. But for the entrepreneur, money is a crucial means of production. Just as the sociologist needs books and free time and the bureaucrat needs arbitrary power, the capitalist needs capital. It makes no more sense to begrudge the entrepreneur his profits than to begrudge the writer or professor his free time and access to libraries and research aides, or the scientist his laboratory and assistants, or the doctor his power to prescribe medicines and perform surgery. Capitalists need capital to fulfill their role in launching and financing enterprise.

The fatal problem of a system without ample personal income and profit is not the lack of incentives but the eclipse of information. Under a system of forced redistribution, the rich, aggressive, and ambitious secure their inevitable advantages not by giving but by taking. They acquire money and power only at the expense of others through political, financial, bureaucratic, and legal maneuvering and manipulation. It is capitalism that best combines the desire to do good and create value with the resources to accomplish these goals.

The conventional wisdom, whether liberal or conservative, regards charity or generosity as essentially simple—just giving things away without calculation or continuing concern with their use. The hero of this narrative is the anonymous donor, while the investor is seen as a Shylock, extorting usurious gains from lending money, or a Scrooge, extracting his profits from the exploitation of workers. A welfare system of direct money grants financed by anonymous taxpayers through the choices of their elected representatives is, in this view, the ultimate expression of compassion and charity.

Dumb money, however, does more harm than good. It is extremely difficult to transfer value to people in a way that actually helps them. Excess welfare hurts its recipients, demoralizing them or reducing them to an addictive dependency that can ruin their lives. The anonymous private donation may be a good thing in itself. It may foster an outgoing and generous spirit. But society as a whole is more likely to become charitable and compassionate if the givers are given unto, if the givers seek some form of voluntary reciprocation. Then the spirit of giving spreads, and wealth gravitates toward those who are most likely to give it back, who are most capable of using it for the benefit of others, who are most knowledgeable and best informed, whose gifts evoke the greatest returns. Even the most indigent families will do better under a system of free enterprise and investment than under a supposedly "compassionate" welfare system that asks no return. The law of reciprocity—that one must supply in order to demand, save in order to invest, consider others in order to serve oneself—is essential for a humane society.

Indeed, it is the genius of capitalism that it recognizes the difficulty of successful giving, understands the hard work and sacrifice entailed in helping one's fellow men, and offers a practical way of living a life of effective charity. True generosity is not soft or sentimental. It consists not of giveaways but of responsible giving. It has little to do with the often lazy or degenerate good works of the gullible, or with all the protest and programs of "social change" and equality urged by the sterile and predatory left.

Capitalism transforms the impulse to give into a disciplined process of creative investment based on a continuous analysis of the needs of others. The investor cannot be fundamentally selfish. A self-centered capitalist is incapable of undertaking the risky but inspired ventures that depend most on an imaginative understanding of the world beyond himself and a generous and purposeful commitment to it.

A new generation of avaricious and egocentric Americans, in the thrall of anti-industrial ideologies, shuns the productive adventures of creation to pursue the comfort and security of the welfare state, abetted by the discount windows at the Federal Reserve, guarantees for "green jobs" and solar enterprises, the triple-A assurances of Fannie and Freddie, and the bonanzas with which we reward litigation against the productive. Right-wing best-sellers, meanwhile, peddle despair, predicting depression and decline. They urge their audience to retreat from business, to buy art, guns, and gold, in a futile search for security in an inevitably insecure world. Self-love keeps even the ambitious and committed from devoting themselves and their wealth to informative experiments on the frontiers of enterprise, where the outcomes will be decided by others acting voluntarily in the market. Instead, they waste their gifts in a quest for power over others, whether in seemingly fail-safe finance protected by government guarantees or in campaigns to halt industrial facilities, extorting funds from real energy companies, or shaking down General Electric again.

Entrepreneurs provide a continuing challenge both to men who refuse a practical engagement in the world, on the grounds that it is too danger-ous or corrupt, and to men who demand power over others in the name of ideology or expertise without first giving and risking themselves and their wealth. Capitalism offers nothing but frustrations and rebuffs to those who, by virtue of their superior intelligence, birth, credentials, or ideals, believe themselves entitled to get without giving, to take without risking, to profit without understanding, and to be exalted without humbling themselves to meet the unruly demands of others in an always perilous and unpredictable life.

It is not surprising, therefore, that the chief source of misunderstand-ing of capitalism is the intelligentsia, who disdain bourgeois or "middle class" values and deny the paramount role of individual enterprise in the progress of the race. Distaste for commercial life is scarcely less common among thinkers on the right than among those on the left. Neither inspired and unruly entrepreneurs nor rambunctious and creative busi-nessmen govern Adam Smith's "great machine," but a crude form of *homo economicus*, motivated by self-aggrandizement.

The error of economists was to found their theory on the mechanism of market exchanges themselves rather than on the business activity that makes them possible and impels their growth. Capitalist exchange indeed works by computations of what might be termed self-interest, as the participants negotiate a price agreeable to each. But this self-interest has nothing to do with avarice; it reflects a mutual transfer of information, allowing an appropriate allocation of resources.

The grander vision of economics fails because it violates a key prin-ciple of information theory. It subordinates a higher and more complex level of activity—the creation of value—to a lower level, its measurement and exchange. In their desire to found a Newtonian science of political economy, generations of economists inflated the instrumental mechanism of trading into a complete economic universe, in which there is little or no room for entropy and surprisal, for the unpredictable activities of free businessmen making entirely new things.

At the heart of capitalist growth, however, is not the mechanistic *homo economicus* but conscious, willful, often altruistic, inventive man. Although a marketplace may work mechanically, an economy is in no sense a great machine. The market provides only the perfunctory dénouement of a tempestuous drama, dominated by the incalculable creativity of entrepreneurs, making purposeful gifts without predetermined returns, launching enterprise into the always unknown future. The market is the conduit, not the content; the low-entropy carrier, not the high-entropy message. Capitalism begins not with exchange but with giving, not with determinist rationality but with creation and surprisal.

The gifts of capitalism generate economic progress chiefly because they constitute an epistemic system, a way of making discoveries and exploiting them. Even the failures in a sense succeed, and the much-remarked "waste" of the system is often redeemed by the accumulation of information and experience, intangible capital held by both the entrepreneurs and the society at large.

Information alone, however, cannot make the system grow. The successful enterprise imparts to the entrepreneur financial resources as well as knowledge. Capitalism is the most effective way of expanding wealth, not because it offers the most powerful incentives—the most tantalizing arrangements of carrots and sticks—but because it links knowledge with power. It gives control over resources and over the future flow of investment not to political bureaucracies of certified experts or to the most avidly self-loving pursuers of leisure and luxury, but to the particular entrepreneurs who manage successful experiments of enterprise. It grants riches to those very individuals who have proved their ability to forgo immediate gratification in pursuit of larger goals, and who refuse to waste or to hedonistically consume their incomes. It assigns further power to those very people, whomever they may be, however unorthodox or un-credentialed, who launch successful projects and commit to them their lives and savings. Under capitalism, economic power flows not to the intellectual, who manipulates ideas and basks in their light, but to the man who gives himself to his ideas and tests them with his own wealth and work.

These often self-denying explorers beyond the bounds of the existing marketplace and its inventories, not some mechanism of exchange, extend the frontiers of human possibility. The greatest damage inflicted by state systems of redistribution and industrial policy is not the "distortion of markets," the "misallocation of resources," or the "dis-coordination" of producers and consumers, but the deflation of capitalist energy, the repression of new entrepreneurial ideas, and the stultification of wealth. Steeply "progressive" tax rates destroy not only incentives; more importantly, they destroy information and preclude entropy. They take from the givers and prevent them from giving again, from reinvesting their winnings in light of the new information that the original gift and its associated experiments generated.

This entrepreneurial process is a method of unveiling surprises. It cannot be captured or codified by a bureaucracy. The economist Jay Forrester, one of the pioneering economic modelers of cities and worlds, has observed that all the libraries of certified knowledge miss the fertile cerebrations, the ebullitions of mind, that make an economy work. Entrepreneurial heads and hands hold an astronomical amount of technical information, skills, intuitions, habits, and practical experience that are often ineffable or unintelligible to anyone who has not pursued the same experimental course. The chemist-philosopher Michael Polanyi has called it "tacit knowing," all the lore and learning that men half-unconsciously collect as they go about their business.

In the move from this multifarious mass of information to written or documented learning alone, we lose an incalculable amount of effective knowledge. Even documented scientific knowledge is not self-evident. In most cases, people outside the field of specialization cannot even read it. As Edward Teller told me at a dinner meeting of the "Coffee Club" in New York decades ago, the problem facing industrial spies trying to steal militarily significant materials is the task of identifying what is important out of the massive flux and welter of scientific papers, engineering schematics, software codes, conference presentations, and computer printouts. Nothing helps them so much as government classification, often color coded, pointing out and prioritizing the prime targets for theft.

Even then, however, the scientific material will generally be revealing only if it contains nothing new. Genuinely entropic science or engineering will often be baffling to analysts who lack knowledge of the scientific context and body of related learning. The cost of appropriating an entropic discovery is mastery of the science that produced it.

In creating effective technical bureaucracies, as Teller pointed out, democratic countries have no advantage over totalitarian ones. Its research and development dominated by the Pentagon bureaucracy, and with most information classified, the United States actually was losing to the Soviet Union after World War II. The Soviets produced nuclear weapons, including the hydrogen bomb, almost in parallel with the United States, and they excelled in the production of missiles. Sputnik gave leadership in space to the U.S.S.R. What saved the United States was information technology, chiefly in computers and microchips in Silicon Valley, in Dallas, and along Route 128, which burst free of the classified and controlled governmental bureaucracies. Although the United States did not prevail in the physics and chemistry of intercontinental missiles, it used microchips and information tools to create MIRVs (multiple independently-targeted reentry vehicles), which could deliver smaller payloads to many targets at once. Later information technologies gave America an insuperable lead not only in armaments but also in industrial growth and creativity.

As Forrester points out, when one moves on from written information to mathematically organized data—the kind of information in a computer model—one suffers a catastrophe of lost learning just as great as the loss incurred by restriction to documented knowledge. Innovation rides on an ever-moving crest of entropic discovery, and by the time the innovations have been fully codified they are often obsolete. Stealing last year's designs from Apple would not let you compete with next year's product in the heads of designers. Yet every socialist plan begins with just such a draconian reduction, just such a conflagration of human learning and skills, leaving only a pile of statistical ashes, a sterile residue of numbers, from which to reconstruct the edifice of economic activity.

No rational determinist scheme can encompass entrepreneurial entropy. Kurt Gödel and Alan Turing defined the problem of the "incomputability" of most innovation. Entrepreneurial entropy begins beyond the boundaries of settled rationality. As a form of new discovery, it passes Gödel's threshold, the point where all logical systems, including mathematics, exhaust their completeness. Entrepreneurship transcends certainty and enters the always-evanescent realm of creation. As the entrepreneur harvests the knowledge from his experimental ventures, it slips away into the past, into history, and into old knowledge—no longer entropic or informative. The entrepreneur is always concerned with the future, riding on the crests of creation.

Because of this problem, no "planned economy" actually observes or fulfills its plan. To avoid disaster, the planners must always remain open to human learning and tacit knowledge, intuition, and experience. But even these forms of information do not suffice. Economies run not only on light, but also on heat and energy; not merely on information, but also on courage and discipline.

Entrepreneurial learning runs deeper than what is taught in schools or acquired in the controlled experiments of social or physical science or gained in the experience of socialist economies. For enterprises are also adventures, with the future livelihood of the investor at stake. He participates with a heightened consciousness and passion and an alertness and diligence that greatly enhance his learning. The experiment may reach its highest possibilities, and its crises and entropic surprises may be exploited to the utmost.

Harvey Leibenstein of Harvard introduced the concept of *x-efficiency* to capture the differences in output between workers, factories, companies, or even nations that use the same measurable inputs.[1] He revealed productivity differences as great as four to one between workers doing the same job in the same plant with the same tools, and as high as 50 percent for plants with apparently identical work forces, tools, and pay. Management, motivation, and spirit—and their effects on willingness to innovate and seek new knowledge—dwarf all measurable inputs in accounting for productive efficiency, for both individuals and groups, for management and labor. A key difference, Leibenstein found, was always in the willingness and ability to transform vague information or hypotheses into working knowledge. In our terms, output is determined above all by openness to the upside surprises and entropy in a falsifiable entrepreneurial experiment.

Without this x-factor, most of the highest possibilities of an economy will remain latent; the lessons of success and failure will come to light only very slowly as the capital of economic knowledge. As Leibenstein's data, everyday observation, and history confirm, ownership is the key to the "spirit" factor. Ownership means exposure to the risks and benefits of productive property, whether it is one's own land and labor or, in a more attenuated form, corporate shares. In a competitive economy in a changing world, the owner lives on the crest of creation, continually informed and inspired, edified and motivated, by the flashes of surprising news about fashion, taste, and technology that can radically shift the profits—the future returns—of what is owned.

Socialism is an insurance policy bought by all the members of a national economy to shield them from risk. But the result is to shield them from knowledge of the real dangers and opportunities in any economic environment. Rather than benefiting from a multiplicity of gifts and experiments, the entire economy absorbs the much greater risk of remaining static in a dynamic world. In a truly capitalist economy, with the risks chiefly borne by the individual citizens and entrepreneurs, and thus vigilantly appraised and treated, the overall system is more stable.

In *Dynamic Economics*, which was an important source for *Wealth and Poverty*, Burton Klein made a valuable distinction between *micro-stability* and *macro-stability*.[2] A government can focus on stabilizing either the microeconomics of the system—the behavior of individual companies and industries—or the macro-level economics of monetary aggregates, gross national product, inflation rates, and currency. Marshalling fiscal and monetary policy, tax rates, and currency values to insure companies and individuals against micro risks like home mortgages, bank deposits, and student loans produces macro-instability. If all the micro risks are insured, the economy becomes inflexible and uncreative, inviting the greatest macro risk: depression.

In capitalism, the entrepreneur, by giving before he takes, pursues a mode of thinking and acting suitable to uncertainty. The socialist makes a national plan in which existing patterns of need and demand are ascertained, and then businesses are contracted to fulfill them; demand comes first. One system is to continuously, endlessly perform experiments, test hypotheses, and discover partial knowledge; the other is to assemble data of inputs and outputs and administer the resulting plans.

Socialism presumes that we already know most of what we need to know to accomplish our national goals. Capitalism is based on the idea that we live in a world of unfathomable complexity, ignorance, and peril and that we cannot possibly prevail over our difficulties without constant efforts of initiative, sympathy, discovery, and love. One system maintains that we can reliably predict and elicit the results we demand—whether it's cost-effective windmill energy or efficient super trains or low-power, high-performance light bulbs. Capitalism asserts that we must give long before we can know what we want and what the universe will return. One system is based on empirically calculable human power; the other on optimism and faith. These are the essential visions that compete in the world and determine our fate.

Seek and you shall find; give and you will be given to. "Supply creates its own demand" is the sequence and the cosmology that distinguishes the free from the socialist economy.

Under capitalism, reason ventures out into a world governed by morality and Providence. The gifts will succeed only to the extent that they are "altruistic," springing from an understanding of the needs of others. They express faith in an essentially fair and responsive humanity. In such a world, a man can venture without the assurance of reward. He can seek the surprises of profit rather than the more limited benefits of contractual pay. He can take initiatives amid radical perils and uncertainties.

When faith dies, so does enterprise. It is impossible to create a system of collective regulation, insurance, and safety that does not finally deaden the moral source of the willingness to face danger and fight, that does not dampen the spontaneous flow of gifts and experiments, that extends the dimensions of the world and the circles of human sympathy.

The ultimate strength and crucial weakness of both capitalism and democracy are their reliance on individual creation. But there is no alternative except mediocrity and stagnation. Demand-based systems can never flourish in a world where events are shaped by millions of human beings, acting unknowably, in fathomless interplay and complexity, in the darkness of time.

The seeker of assurance and certainty lives always in the past, which alone is sure, and his policies, despite the "progressive" rhetoric, are necessarily reactionary. Certain knowledge, to the extent that it ever comes, is given us only after the moment of opportunity has passed. The venturer who awaits the emergence of a safe market, the tax-cutter who demands full assurance of new revenue, the leader who seeks a settled public opinion, will always act too timidly and too late. The future is available only to entrepreneurs, on the crest of creation, seeking the entropy of giving.

Acknowledgments

THIS BOOK IS the product of multiple inputs—a fan-in of several streams of ideas and editorial guidance aimed to ignite a fan-out of *Knowledge and Power*. So crucial were my two early editors, Richard Vigilante and Louisa Gilder, in shaping this book that both will receive royalties, even if I must reserve the credit, blame, legal liability, and residual claims in the usual way to myself as author.

Not receiving royalties but deserving them also is Bruce Chapman, my old friend and colleague since college, who did a close edit of the book before Tom Spence of Regnery took over. Spence produced the kind of tenacious and detailed editorial hand allegedly no longer available at mainstream publishers.

Many of the ideas in this book are an ebullition from the unique crucible of Discovery Institute in Seattle, under the leadership of Chapman and his able successor Steve Buri, with pivotal guidance from long-time chairman Tom Alberg of Madrona Capital. Gathered at Discovery is the world's leading faculty of scholars on information theory and biology. Among them are Douglas Axe, Michael Behe (*Darwin's Black Box*), David Berlinski (*The Ascent of the Algorithm*, et al.), William Dembski (*No Free Lunch*), Michael Denton (*Nature's Destiny*), Stephen Meyer (*The Signature in the Cell, Darwin's Doubt*), Jay Richards (*The Privileged Planet*, with Guillermo Gonzalez), Richard Sternberg (*The Metaphysics of DNA*, forthcoming), and Jonathan Wells (*The Myth of Junk DNA*). Other important contributors to my thinking from outside Discovery are Steven Waite (*Quantum Investing*), Tom Bethell (*The Noblest Triumph: Property and Prosperity Through the Ages*), and James Cheney, keeping me up on ideas and developments in Israel.

Manifesting the value of the information theory path in biology (and other disciplines involving humans) is a company called Immusoft, launched by Matt Scholz, manager for twelve years of information technology at Discovery. As this book maintains, pharmacology has long suffered from the materialist superstition: the view that the human body is best understood as a chiefly physical and chemical phenomenon. This assumption leads to a pharmaceutical agenda of fighting disease by injecting chemical compounds into the bloodstreams of sick people or feeding them these compounds as pills or capsules. But the body contains trillions of ribosomes that can produce proteins to order at the time and location they are needed.

Immusoft aims to use this resource to turn people's cells into drug factories. It is based on the insight that the body is largely governed by information flows, originating in DNA, and thus is a programmable platform, like other information systems. Financed in part by Peter Thiel, venture capitalist supreme, Immusoft currently uses viruses (and will eventually use synthetic vectors) to program the stem cells or plasmid B-cells in the bone marrow organically to produce the compounds needed by the body—24/7, if necessary, for a lifetime—for chronic conditions. Immusoft's method has been tested successfully on animals, and the company provides an existence proof of the promise of the Discovery approach to information and biology that I describe in this book.

The human body and the economy are information systems and intelligible through information theory. So is the universe. Ultimately it is families that organize most of the information that animates life. My own family was vital to the writing of this book, edited by Louisa and supported in a critical writing stint in Thailand by my refugee-camp malaria-project doctor daughter Mellie. Encouragement was always forthcoming when needed from Richard and Nannina, and the Lenox Coffee Shop supplied the necessary caffeine.

All my activities revolve around my wife Nini, who gives them meaning, substance, and soul.

—George Gilder

KEY TERMS
FOR THE NEW
ECONOMICS

A TENDENTIOUS GLOSSARY

Information theory: Based on the mathematical ideas of Claude Shannon and Alan Turing, an evolving discipline that depicts human creations and communications as transmissions across a channel, whether a wire or the world, in the face of the power of noise, with the outcome measured by its "news" or surprise, defined as "entropy" and consummated as knowledge.

Since these creations and communications can be business plans or experiments, information theory provides the foundation for an economics driven not by equilibrium or order but by falsifiable entrepreneurial surprises.

Information theory enables and describes our digital world.

Shannon entropy: Information measured by surprisal or unexpected bits. The entropy of a message represents the freedom of choice and the room for creativity of the message's composer. The larger the set of possible messages or the larger the alphabet of symbols, the greater the composer's choice and the higher the entropy or uncertainty resolved by the message—and thus the greater the information.

Entropy is most simply measured by the number of binary digits needed to encode the message, and it is calculated as the sum of the base 2 logarithms of the probabilities of the message's components. (The logarithms of probabilities between one and zero are always negative quantities; entropy is rendered positive by a minus sign in front of the sum.[1]) Order is predictable and thus low-information and low-entropy.

Boltzmann's or physical entropy: Heat (the total energy of all molecules in a closed system) over temperature (the average energy of the individual molecules).

In a system divided between hot and cold entities, Boltzmann's entropy begins as zero—when we know most about the arrangement of the system—and it reaches maximum entropy when those hot and cold entities merge and we know least. Boltzmann therefore identified this number with missing information, or uncertainty about the arrangement of the molecules, opening the way for Shannon and information theory. "Boltzmann hurled the concept of information into the realm of physics," as Hans Christian von Baeyer writes.[2]

In a closed system, physical entropy always increases. Heat flowing to the colder object reduces the available energy of potential differences. The system converges to the average temperature. Entropy thus imparts a dimension of statistically irreversible time to physical events.

Although the Boltzmann and Shannon equations are similar, Boltzmann's entropy is analog and governed by the natural logarithm

---

[1] This negative sign in Shannon's equation prompted some eminent theorists—Norbert Wiener and Erwin Schrödinger, among others—to blunder into the idea of negentropy, which is an oxymoron. "Because all probabilities must range from zero to one the logarithm is negative and that means that [entropy] is positive," wrote the great Hubert Yockey. The temptation of negative entropy comes from a confusion of entropy and complexity with order.

[2] Hans Christian van Bayer, *Information: The New Language of Science* (Cambridge: Harvard University Press, 2005).

"e." Shannon entropy is digital and governed by the binary logarithm, log base 2.

Gödel's incompleteness theorem: Kurt Gödel's discovery in mathematical logic, published in 1931, that any formal system powerful enough to express the truths of arithmetic will be incomplete and dependent on axioms not reducible to the system—truths that cannot be proved within the system itself.

In developing his proof of incompleteness, Gödel invented a mathematical "machine" that used numbers to encode and prove axioms also represented by numbers. Thus he anticipated the central discoveries of computer science as pioneered by Alan Turing and John von Neumann.

By showing that mathematics could not be hermetically sealed or exhaustively consistent and deterministic, Gödel opened the way to postmodern mathematics: a combinatorial mathematics of software, life, and creativity replacing the deterministic mathematics of physics.

Turing machine: An abstract universal computer model, devised in 1936 and inspired by Gödel's proof, consisting of a control unit administering a set of instructions and reading, writing, erasing, and moving one space at a time back and forth along an infinitely long tape divided into squares along its length. Turing proved that this simple hypothetical machine could perform any computable function, and, alas, that most numbers are not computable. Turing also showed that a universal computer could not calculate whether any particular program would ever halt.

Turing's machine was a general-purpose computer because it commanded infinite time and space. Real-world computers are necessarily restricted to finite sets of applications. But by demonstrating the limits of computer logic, Turing opened the way to creativity in software, giving the programmer the role of "oracle."

Shannon carrier or channel: The conduit or path along which the signals flow.

For the content to be differentiated from the carrier at the receiver, the carrier must be low-entropy. Thus it takes a low-entropy carrier to bear high-entropy creations. It takes a meaningless medium to carry important meanings.

The most reliable low-entropy carrier is the electromagnetic spectrum, which is governed everywhere by the absolute speed of light.

Shannon's channel capacity formula shows that, in the face of noise (see the following definition), increasing the bandwidth of the channel is more efficient than increasing the power of the signal. An increase of the channel's bandwidth produces a commensurate linear increase in the flow of information. Increasing the power of the signal expands information flow only by the logarithm of the increase. So to increase the capacity of a channel, Shannon counsels a strategy of "wide and weak"—expanded bandwidth at low power.

Noise: Any influence of the conduit on the content; an undesired disturbance in a communications channel.

Noise in information theory is analogous to ohmic resistance in electronics. This analogy implies that noise is also crucial to signal and to entropy. The ideal is called "white noise" (analogous to white light), which contains all frequencies not serially correlated and thus independent in time. Since the impulses of white noise are uncorrelated, it shows the highest entropy: no impulse is predictable from its precursor. A stream of random noise is in principle indistinguishable from a series of creative surprises.

White noise is a hypothetical ideal because if it really contained all frequencies it would entail infinite energy. In practice, all noise dwindles away at some high-energy frequencies. A more usual spectrum of noise is "brown" noise, characterizing the spontaneous Brownian thermal motion of molecules first mathematically defined by Einstein.

Chaitin's Law: Static, eternal, perfect mathematics cannot model dynamic creative life. Gregory Chaitin of Bell Labs, the inventor of algorithmic information theory, demonstrated the need to transcend the

Newtonian mathematics of physics and adopt post-modern mathematics—the mathematics that follows Gödel, Turing, and von Neumann, "the mathematics of creativity."

Knowledge: Verified or falsified information, processed and interpreted into meaning and significance, wisdom and leadership. Usually local, particular, and derived from experience, knowledge is the source of value in an economy.

Matter is conserved; knowledge, by contrast, expands and accumulates. All economic transactions consist of the trading of the transactors' differential knowledge.

Power: The ability to endow knowledge with effectiveness, providing reliable low-entropy channels of law, finance, and governance, or to suppress or overrule knowledge by coercion.

Economies prosper to the extent that knowledge, intrinsically dispersed through the system in the minds of individuals, is complemented by a similar dispersal of power. Free markets and uncontrolled prices typically achieve this goal. Combinations of government and business frustrate it.

In economics as in communications, Shannon's strategy of "wide and weak" applies. In the face of noise, expand the bandwidth of knowledge and disperse the power.

Notes

CHAPTER ONE: THE NEED FOR A NEW ECONOMICS

This opening chapter is based on a draft by Richard Vigilante, my virtual "editor-in-chief" for all my books since 1982, with whom I worked out the structure of this work over many weeks. The chapter defines the element missing in all contemporary economic models: the unexpected boons of creativity. As Princeton economist Albert Hirshman wrote in an early issue of the *Public Interest*, creativity always comes as a surprise to us. If it didn't, we would not need it; we could plan it, and socialism would work. Surprise is not exogenous to capitalism; it is its essence, and the essence of information and enterprise.

1. Steven Levitt and Stephen Dubner, *Freakonomics: A Rogue Economist Explains the Hidden Side of Everything*, revised and expanded edition, (William Morrow, 2006); and Nobel Laureate Gary Becker with Guity Nashat Becker, *The Economics of Life: From Baseball to Affirmative Action to Immigration, How Real-World Issues*

Affect our Everyday Life (New York: McGraw-Hill, 1996). In *Redeeming Economics: Rediscovering the Missing Element* (Wilmington, DE: ISI Books, 2010), John D. Mueller debunks some of Levitt's and Becker's more facile crowd-pleasers, providing a different take on the need for a new economics. See especially Chapter 8, "An Empirical Test: Fatherhood and Homicide," 185–187.

2. Mark Skousen, *Vienna & Chicago: A Tale of Two Schools of Free Market Economics* (Washington, D.C.: Capital Press, 2005), is a masterpiece of pithy and learned economic history and analysis covering these canonical sources of Austrian and Chicago economic thought. Also insightful and edifying is Skousen, *The Big Three in Economics: Adam Smith, Karl Marx, and John Maynard Keynes* (Armonk, NY: M.E. Sharpe, 2007). Valuable background for this book is David Warsh, *Knowledge and the Wealth of Nations* (New York: W.W. Norton & Company, 2007), and G.L.S. Shackle, *Epistemics and Economics: A Critique of Economic Doctrines* (Cambridge, UK: Cambridge University Press, 1972).

3. John Allison, *The Financial Crisis and the Free Market Cure* (New York: McGraw-Hill, 2012). Allison tells the story of how his bank thrived through the crash. He also offers a needed corrective for the lionization of the "big bankers," such as Hank Paulson and Tim Geithner.

4. The definitive statement of the knowledge problem comes in Friedrich Hayek, "The Use of Knowledge in Society," *American Economic Review* XXXV (1945): 30. "The peculiar character of the problem of a rational economic order is determined precisely by the fact that the knowledge of the circumstances of which we must make use never exists in concentrated or integrated form but solely as the dispersed bits of incomplete and frequently contradictory knowledge which all the separate individuals possess. The economic problem of society is thus not merely a problem of how to allocate 'given' resources—if 'given' is taken to mean given to a single mind which deliberately solves the problem set by these 'data.' It is rather a problem of how to secure the best use of resources known to any of the members of society, for ends whose relative importance only

these individuals know. Or, to put it briefly, it is a problem of the utilization of knowledge which is not given to anyone in its totality.... The problem is thus in no way solved if we can show that all the facts, *if* they were known to a single mind (as we hypothetically assume them to be given to the observing economist), would uniquely determine the solution; instead we must show how a solution is produced by the interactions of people each of whom possesses only partial knowledge. To assume all the knowledge to be given to a single mind in the same manner in which we assume it to be given to us as the explaining economists is to assume the problem away and to disregard everything that is important and significant in the real world."

5.   Thomas Sowell, *Knowledge and Decisions* (New York: Basic Books, 1980), introduced me to the proliferating ramifications of the subject of knowledge in economics and the damage inflicted by the intrusions of political power. His book could have been titled "Knowledge and Power," but the source of the title for my book was Arnold Kling, *Unchecked and Unbalanced: How the Discrepancy Between Knowledge and Power Caused the Financial Crisis and Threatens Democracy* (Lanham, MD: Rowman & Littlefield, 2010), which begins: "This book represents an attempt to explore the problem of the discrepancy between the trends in two phenomena: *knowledge* is becoming more diffuse, while political *power* is becoming more concentrated." My book shows that Kling's insight finds deep roots in the information theory that underlies the modern world economy.

CHAPTER TWO: THE SIGNAL IN THE NOISE

In this chapter, the surprise is Qualcomm corporation, which self-consciously introduced information theory to American commerce. In a determinist economic model, entrepreneurship will seem to be noise in the channel. Qualcomm used such noise to bear information. My daughter Louisa Gilder, author of *The Age of Entanglement* (Knopf, 2008), shaped and edited this and the subsequent two chapters.

1. John R. Pierce, *An Introduction to Information Theory: Symbols, Signals, and Noise*, second revised edition (New York: Dover Books, 1980). This wide-ranging text by one of Claude Shannon's closest colleagues and disciples proceeds from physics to music and illustrates the power of information theory to illuminate all intellectual life.

2. James Gleick, *The Information* (New York: Pantheon, 2011). The most comprehensive and comprehensible study of information theory, enlivened with an extensive biography of Claude Shannon.

3. Pierce, *An Introduction to Information Theory*, 178.

4. Pierce, *An Introduction to Information Theory*. The deeper permutations of noise are well explained in Edward Beltrami, *What Is Random?: Chance and Order in Mathematics and Life* (New York: Springer-Verlag, 1999), 131–143, where he expounds on a spectrum of white, pink, red, and black noise.

CHAPTER THREE: THE SCIENCE OF INFORMATION

Such is the sequestration of disciplines in the American Academy that most economists and social scientists, as well as biologists and even physicists, have no idea of who Claude Shannon was. I hope this chapter remedies the neglect of one of the paramount thinkers of the information era. Better known is Shannon's colleague in cryptography during World War II and co-founder of information theory after it, Alan Turing, who might possibly have rivaled Shannon in creative extensions of computer science. But following the war, Alan Turing committed suicide by eating a poisoned apple after having undergone court-mandated estrogen therapy to rein in his public homosexuality. The Apple logo, with its missing bite, is thought by some to be an homage to Turing, but Steve Jobs said he only wished he had been that smart.

1. Shannon's works have been collected in an IEEE Tome: Claude Shannon, *Collected Papers* (New York: Wiley-IEEE Press, 1993), which also include several interviews with Shannon and his colleagues. Other biographical details come from James Gleick, *The*

Information; John Horgan, *IEEE Spectrum* (April 1992); and *Scientific American* (January, 1990): 20–22.

2. Seth Lloyd, *Programming the Universe* (Knopf, 2006). An essay on the information theory of physics by a leading MIT physicist and developer of the conceptual foundations of quantum computing.

3. Peter Drucker, *Innovation and Entrepreneurship* (New York: HarperCollins, 1985).

CHAPTER FOUR: ENTROPY ECONOMICS

The title of this chapter was coined by Bret Swanson for his estimable blog of Telecom analysis and information theory. His work can be found online at http://entropyeconomics.com/index.php/about/bret-swanson/.

1. For these purposes, I take information and knowledge as kindred terms (knowledge being processed information). Peter Drucker, "The Knowledge Age," in *The Age of Discontinuity* (New York: Harper and Row, 1968). Drucker sustained his pioneering preoccupation with the ascendant role of knowledge throughout his career.

2. Daniel Bell, *The Coming of Post Industrial Society* (New York: Basic Books, 1973), wrote authoritatively of the increasing role of knowledge and information in a "creativity culture," though he underestimated entrepreneurial skills in his credulous stress on professional expertise, and in his academic itch to redeem the idiom of Marxism.

3. Among many epigrams on the media, information, and knowledge trinities, Marshall McLuhan spoke presciently in 1967 on "The Best of Ideas," CBC radio: "One of the effects of living with electric information is that we live habitually in a state of information overload. There's always more than you can cope with."

4. Alvin Toffler, *The Third Wave* (New York: Bantam, 1980) was the most detailed and articulate exposition of the move toward an information society and a knowledge economy. Selling millions of

copies around the globe, it may have been the most prophetic book of the age, though intellectuals characteristically disdained it for shallowness and metaphorical aerobatics.

5. Stewart Brand, author of *The Whole Earth Catalog* and of the mantra "information wants to be free," was a pioneer of information-era companies and concepts. A swashbuckling figure who lives on a tugboat in Sausalito Harbor, Brand was immortalized by Tom Wolfe in *The Electric Kool-Aid Acid Test*.

6. John Perry Barlow is the poet of the information age, known for his contributions to the ardent infoscapes of the Grateful Dead ("I Need A Miracle" and "Estimated Prophet," among many other canonical non-Robert Hunter lyrics). He also wrote incandescent tributes to the information and knowledge economy and was the author of "The Declaration of the Independence of Cyberspace": "Governments of the Industrial World, you weary giants of flesh and steel, I come from Cyberspace, the new home of Mind. On behalf of the future I ask you of the past to leave us alone. You are not welcome among us. You have no sovereignty where we gather.... Governments derive their just powers from the consent of the governed. You have neither solicited nor received ours. We did not invite you. You do not know us, nor do you know our world. Cyberspace does not lie within your borders."

7. Shannon, "*Omni* interview" in *Collected Papers*.

8. Walter Isaacson, *Steve Jobs* (New York: Simon and Schuster, 2011).

9. This Henry Ford quote is said to be apocryphal, but it is always ascribed to him and it captures the essence of his approach to the market.

CHAPTER FIVE: ROMNEY, BAIN, AND THE CURVE OF LEARNING

This chapter was written in the expectation that Mitt Romney would be President by now. Many a slip.... But the story of Bain and Company and the learning curve remain central to this book.

1. Michael Kranish and Scott Helman, *The Real Romney* (New York: HarperCollins, 2012).

2. Michael Jensen, "The Modern Industrial Revolution, Exit, and the Failure of Internal Control Systems," *The Journal of Finance* 48, no. 3 (July 1993): 831–880. Jensen delivered this paper as a keynote to the 53rd Annual Meeting of the American Finance Association in Anaheim, CA, January 5–7, 1993. Later he and his students updated the numbers.

3. Mark Skousen, *The Structure of Production*, paperback edition with new introduction (1990; reprint, New York: NYU Press, 2007).

4. Michael Rothchild, *Bionomics: Economy as Ecosystem*, reissue edition (Hammond, IN: Owl Books, 1995).

5. Raymond Kurzweil, *The Singularity is Near* (New York: Penguin Books, 2006). A classic of technological analysis and projection, with elegantly written "dialogues" at the end of each chapter, it showed that America's inventor-futurist is also a superb writer. His assumption that the mind is essentially a material computational machine, extended in a new seminal work, *How to Build a Mind* (New York: Viking, 2012), fails to persuade me, while as usual teaching me much. I just do not believe that transistors or on-off switches, however many and however fast, can transcend their programming to become conscious. Kurzweil's book on the mind remains the most fascinating explanation of how to build a natural language speech recognizer.

6. Henry Adams, *The Education of Henry Adams* (Library of American Paperback Classics, 2009). In this autobiographical meditation written at the end of the nineteenth century, Adams dubs his insight into the dynamics of economic growth "The Law of Acceleration," and limns out a curve for energy usage that anticipates the scores of similar exponentials in Kurzweil's work.

7. Rand Paul, "Filibuster on U.S. Senate Floor" (speech, Washington, D.C., March 6, 2013), Cato Institute, http://www.cato.org/publications/commentary/john-brennan-won-did-meaning-america-survive. "When we passed Obamacare, it was 2,000-some-odd pages. There have been 9,000 pages of regulations written since. Obamacare had 1,800 references that the Secretary of Health shall decide

at a later date. We (the people) gave up that power. We gave up power that should have been ours, that should have been written into the legislation. We gave up that power to the executive branch ... many of whom we call bureaucrats, unelected."

8.   John Goodman, *Priceless: Curing the Healthcare Crisis* (Oakland, CA: The Independent Institute, 2012). Also, Arnold Kling, *Crisis of Abundance: Rethinking How We Pay for Health Care* (Washington, D.C.: Cato Institute, 2006).

CHAPTER SIX: THE EXTENT OF LEARNING

In 2010, in *The Rational Optimist* (Harper, 2010), an epochal achievement in the history of economic thought, Matt Ridley, in different terms, provided a crucial insight for this chapter: "The cumulative accretion of knowledge by specialists that allows us each to consume more and more different things by each producing fewer and fewer is, I submit, the central story of humanity." Describing "prosperity" as "exchange and specialization—more like the multiplication of labor than the division of labor," he essentially depicts this process as the transmutation of knowledge and coal, ideas and energy, into "the 'creation' of time," which enabled the abolition of slavery and the emergence of the modern world. I cannot improve on Ridley's argument; I can only point to his book in admiration and offer a few embellishments from information theory, treating the "accretion of knowledge" as a process of learning.

In an information economy, learning by experiment is the crucial force of progress. Falsifiability is the test of useful hypotheses of enterprise. So the extent of learning replaces the extent of the market as the measure of creativity in the division of labor. For more detailed and technical analysis of the processes of learning, Yaser S. Abu-Mostafa, Malik Magdon-Ismail, and Hsuan-Tien Lin, all trained at Caltech, offer the definitive *Learning from Data: A Short Course* (AMLbook.com, 2012). Jeffrey L. Funk, *Technology Change and the Rise of New Industries* (Stanford, CA: Stanford Business Books, 2013) definitively examines, with many edifying examples, the futility of demand-side approaches to innovation.

1.　George J. Stigler, "The Successes and Failures of Professor Smith," (speech presented at a meeting of the Mont Pelerin Society, St. Andrews, Scotland, 1976), The University of Chicago Graduate School of Business, http://www.chicagobooth.edu/research/select edpapers/sp50c.pdf. The question is, "If scale economies are so important, how do small firms manage to exist at all? How do we get the sort of competition essential to the Invisible Hand?" (Warsh, *Knowledge and the Wealth of Nations*, 46). As early as 1951, George Stigler was already wrestling with this issue: "Either the division of labor is limited by the extent of the market, and, characteristically, industries are monopolized; or industries are characteristically competitive, and the [Invisible Hand] theorem is false or of little significance." Stigler, "The Successes and Failures of Professor Smith," 46.

2.　Harold J. Berman, *Law and Revolution* (Cambridge: Harvard University Press, 1983), 340. This definitive tome traces the sources of Western law to the Papal Revolution of the twelfth century, when the Roman Catholic Church gained its political and legal independence from the political establishments of kings and feudal lords. The mercantile systems stemmed from this original crucible of the law.

3.　Gordon Moore ascribed his "law" to Carver Mead in an account of the history of the law on its thirtieth anniversary. The source of the Moore's Law analysis, and much of the material in this chapter, is my book *Microcosm: The Quantum Era in Economics and Technology* (New York: Simon and Schuster, 1989).

4.　Chris Anderson, *Makers: The New Industrial Revolution* (New York: Crown Business, 2012). Anderson lucidly expounds the idea of a new desktop revolution, following desktop computing and desktop publishing, enabling desktop manufacture. An enthusiast who has begun his own company in the field, making do-it-yourself drones, he may well exaggerate the pace of the change, as some of us enthusiasts tend to do. But as with his previous book, *The Long Tail* (New York: Hyperion, 2008), he puts his finger on the implications of a radical development of his epoch, a new long tail of customized manufacturing by neighborhood artisans. From the ribosome in biology to the 3D printer, programmed atoms and molecules rule. As Gregory Chaitin points out in *Proving Darwin: Making Biology Mathematical* (New York: Vintage, 2012), all these long tails derive from the promethean mind of John von Neumann, whose Gödelian math and architecture underlie the conventional computer and whose cellular automata provided the math behind the 3-D printer.

CHAPTER SEVEN: THE LIGHT DAWNS

David Warsh introduced me to William Norhaus's Promethean paper on the compound learning curves of illumination. This essay shows that Moore's Law is not an anomaly or a special effect of digital computing but a characteristic manifestation of technological advance based on learning and information.

1. Robert Solow, "A Contribution to the Theory of Economic Growth," *The Quarterly Journal of Economics* 70, no. 1 (MIT Press, 1956). This paper is the starting point for David Warsh's *Knowledge and the Wealth of Nations* (New York: W. W. Norton, 2006), which in turn is the inspiration for this chapter and other chapters.

2. Robert J. Gordon, "Is U.S. Economic Growth Over? Faltering Innovation Confronts the Six Headwinds" (Working Paper no. 18315, National Bureau of Economic Research, August 2012): "This paper raises basic questions about the process of economic growth. It questions the assumption, nearly universal since Solow's seminal contributions of the 1950s, that economic growth is a continuous process that will persist forever. There was virtually no growth before 1750, and thus there is no guarantee that growth will continue indefinitely. Rather, the paper suggests that the rapid progress made over the past 250 years could well turn out to be a unique episode in human history."

3. Warsh, *Knowledge and the Wealth of Nations*.

4.　William D. Nordhaus, "Do Real-Output and Real-Wage Measures Capture Reality? The History of Lighting Suggests Not," Cowles Foundation for Research in Economics at Yale University (New Haven: 1998). The epochal paper was delivered first to the National Bureau of Economic Research in 1993.

5.　Warsh, *Knowledge and the Wealth of Nations*, 336. For a more detailed and authoritative account of the Nordhaus revelation, see Chapter 24: "A Short History of the Cost of Lighting."

6.　Eric Savitz and Bill Watkins, "How Silicon Will Spur A Boom In Solid-State Lighting," *Forbes*, November 23, 2012, http://www. forbes.com/sites/ciocentral/2012/11/23/how-silicon-will-spur-a-boom-in-solid-state-lighting/.

CHAPTER EIGHT: KEYNES ECLIPSES INFORMATION

In researching this chapter, I rediscovered the rich and comprehensive economic writings of Mark Skousen, who has told the story of the overthrow of Keynesianism better than anyone else, probably because he contributed so trenchantly and relentlessly to the argument over two decades. Skousen's work on the structure of production subsumes and transcends Eugen von Bohm-Bawerk's original Austrian analysis in *The Positive Theory of Capital* (South Holland, IL: Libertarian Press, 1959).

1.　James Livingston, *Against Thrift: Why Consumer Culture Is Good for the Economy, the Environment, and Your Soul* (New York: Basic Books, 2011).

2.　Paul Krugman, *The Return of Depression Economics* (New York: W. W. Norton, 2009) and *End This Depression Now* (New York: W. W. Norton, 2012).

3.　Krugman, *End This Depression Now.*

4.　James Livingston, *Against Thrift.*

5.　Livingston, "It's Consumer Spending, Stupid!," *New York Times*, October 25, 2011.

6.　Livinston, "It's Consumer Spending, Stupid!"

7.　Mark Skousen, *The Structure of Production* (New York: NYU Press, 2007). An accessible summary of Skousen's "Aggregate

Production Structure" time-oriented model appears in Skousen, *Economic Logic*, revised third edition (Washington, D.C.: Capital Press, 2010), 59–66.

8. Skousen, *Economic Logic*.

9. Yi Wen, "Savings and Growth under Borrowing Constraints: Explaining the "High Savings Rate' Puzzle", (Working Paper, The Federal Reserve Bank of St. Louis, June 2010). The paper explains: "A leading alternative view is that the PIH [Milton Friedman's famous Permanent Income Hypothesis which assumes a propensity to save governed by lifetime savings targets rather than opportunity costs] fails because it is based on, among other things, the assumption of exogenous rates of returns to financial assets (i.e., the real interest rate). In a production economy with productive assets (such as capital), the real rates of return are determined by the marginal products of such assets. When asset returns are so determined, they will respond to changes in productivity growth, which is the fundamental source of changes in permanent income. A permanent increase in total factor productivity (TFP) raises the rate of return to capital, so investment demand will increase, resulting in a higher equilibrium saving rate through a higher real interest rate. Consequently, in contrast to the prediction of the PIH [permanent income hypothesis], standard general-equilibrium growth theory suggests that household saving may increase rather than decrease in response to a higher permanent income (as implied by the analysis of Chen, Imrohoroglu, and Imrohoroglu, 2006)." A pithy summary of the argument is found in E. Katarina Vermann, "Wait, Is Saving Good or Bad? The Paradox of Thrift," Federal Reserve Bank of St. Louis *Page One Economics Newsletter*.

10. Yi Wen, "Savings and Growth."

11. Skousen, letter to the *New York Times*, personal communication.

12. Skousen, *Economic Logic*.

13. Richard Posner, *The Crisis of Capitalist Democracy* (Cambridge: Harvard University Press, 2010) and *A Failure of Capitalism* (Cambridge: Harvard University Press, 2009). The 2010 book was apparently rushed into print to mitigate the excessive leftist credulity

exhibited by the 2009 crisis volume, replete with such distractions as an alleged "climate change" crisis displacing the early "global warming." (Posner could be blind to the comedy of "climate change" casuistry only as long as he believed he retained a chance for ascent to the Supreme Court, where admittedly he belongs).

14. Jean-Baptiste Say, *A Treatise on Political Economy* (1855; reprint, Philadelphia: Lippincott, Grambo & Co., 1803), available online at www.econolib.org/library/Say/sayT15.html. "The same principle leads to the conclusion, that the encouragement of mere consumption is no benefit to commerce; for the difficulty lies in supplying the means, not in stimulating the desire of consumption; and we have seen that production alone, furnishes those means. Thus, it is the aim of good government to stimulate production, of bad government to encourage consumption." A judicious presentation of the debate appears in Steven Kates, *Say's Law and the Keynesian Revolution: How Macroeconomic Theory Lost its Way* (Northampton, MA: Edward Elgar, 1998).

15. I owe this circular-flow argument, among many others in this book, to Richard Vigilante of Whitebox Investment Advisors, Minneapolis, MN.

16. Mark Collette, "The Wildcatter: Corpus Christi's Gregg Robertson, key member of Eagle Ford discovery, named 2012 Newsmaker of the Year," *Corpus Christi Caller-Times*, December 29, 2012, www.caller.com/news/2012/dec/29/the-wildcatter-corpus-christis-gregg-robertson/?partner=RSS.

CHAPTER NINE: FALLACIES OF ENTROPY AND ORDER

Howard Bloom is author of his own luminous theory of economics, *The Genius of the Beast: A Radical Re-vision of Capitalism* (Amherst, NY: Prometheus, 2010). He stresses meaning (the cosmos as a meaning machine) and disparages Shannon's information theory for leaving out meaning. Shannon focuses on the channels or conduits of meaning rather than on meaning itself (the content of the channel). Thus he offers a theory that can accommodate all kinds of meanings (and creations) just as DNA could accommodate all forms of creatures through all time.

Gregory Chaitin, author of algorithmic information theory, makes the point in his iconoclastic *Proving Darwin: Making Biology Mathematical* (New York: Vintage Books, 2013), "Life is plastic, creative! How can we build this out of static, eternal, perfect mathematics? We shall use postmodern math, the mathematics that comes after Godel, 1931, Turing, 1936, open not closed math, the math of creativity...."

1. Howard Bloom, "Heresy Number Three: Prepare to be Burned at the Stake (The Second Law of Thermodynamics—Why Entropy is an Outrage)," in *The God Problem: How a Godless Cosmos Creates* (New York: Prometheus Books, 2012).

2. Howard Bloom, *The God Problem*.

3. Howard Bloom, *The God Problem*.

4. Hubert Yockey, *Information Theory, Evolution, and the Origin of Life* (Cambridge, UK: Cambridge University Press, 2005). This lucid, magisterial treatment of the subject is full of the author's well-seasoned opinions that began with the first conference on information theory and biology in Gatlinburg, Maryland, in 1954, where he introduced George Gamow and his pioneering concept of a code carried in DNA. The 2005 text follows Yockey's definitive *Information Theory and Molecular Biology* (Cambridge, UK: Cambridge University Press, 1992). For a less mathematical exposition of a key part of the argument, see Dean L. Overman, *A Case Against Accident and Self-Organization* (Lanham, MD: Rowman & Littlefield, 1997).

5. Paul Krugman, *The Self-Organizing Economy* (New York: Wiley-Blackwell, 1996).

6. Friedrich Hayek, *The Constitution of Liberty* (Chicago: University of Chicago Press, 1978). "[T]he ultimate aim of freedom is the enlargement of those capacities in which man surpasses his ancestors and to which each generation must endeavor to add its share— its share in the growth of knowledge and the gradual advance of moral and aesthetic beliefs, where no superior must be allowed to enforce one set of views of what is right or good and where only further experience can decide what should prevail."

CHAPTER TEN: ROMER'S RECIPES AND THEIR LIMITS

My daughter Louisa Gilder persuaded me to include this chapter, which presents the work of the most informative and Shannonesque of contemporary economists. She also researched and drafted much of it. But according to the usual convention, the errors and conclusions belong to Dad.

1.   David Warsh, *Knowledge and the Wealth of Nations* (New York: Norton, 2007), 376.
2.   Arnold Kling and Nick Schulz, *Invisible Wealth* (New York: Encounter Books, 2011), 80.
3.   Kevin Kelly, "The New Economics of Growth," *Wired*, June 1996.
4.   Paul Romer, "New Goods, Old Theory, and the Welfare Costs f Trade Restrictions" (Working Paper no. 4452, National Bureau of Economic Research, September 1993).
5.   Russell Roberts, "An Interview with Paul Romer on Economic Growth," *Library of Economics and Liberty*, November 5, 2007, http://www.econlib.org/library/Columns/y2007/Romergrowth.html#.
6.   Roberts, "An Interview with Paul Romer."
7.   Roberts, "An Interview with Paul Romer."
8.   Ronald Bailey, "Post-Scarcity Prophet," *Reason*, December 1, 2001, http://reason.com/archives/2001/12/01/post-scarcity-prophet/print.
9.   Bailey, "Post-Scarcity Prophet."
10.   Bailey, "Post-Scarcity Prophet."
11.   Bailey, "Post-Scarcity Prophet."
12.   Henry Adams, *The Education of Henry Adams*.
13.   Nicolas Negroponte, *Being Digital* (New York: A. A. Knopf, 1999).
14.   Thomas Jefferson, "Letter to Isaac McPherson," August 13, 1813, available online at http://press-pubs.uchicago.edu/founders/documents/a1_8_8s12.html.
15.   Warsh, *Knowledge and the Wealth of Nations*, 378.
16.   Bailey, "Post-Scarcity Prophet."
17.   Roberts, "An Interview with Paul Romer."

18. Bailey, "Post-Scarcity Prophet."
19. Roberts, "An Interview with Paul Romer."
20. "25 Most Influential Americans," *TIME*, April 21, 1997, available online at http://www.time.com/time/magazine/article/ 0,9171,137548,00.html.
21. Steven Pearlstein, "An Economist, An Academic Puzzle, and a Lot of Promise," *Washington Post*, May 8, 2009, http://articles.wash ingtonpost.com/2009-05-08/opinions/36822397_1_college-stu dents-textbooks-economics.
22. Kling, *Invisible Wealth*, 90–91.
23. Sebastian Mallaby, "The Politically Incorrect Guide to Ending Poverty," *Atlantic*, July/August 2010, http://www.theatlantic.com/ magazine/archive/2010/07/the-politically-incorrect-guide-to-ending-poverty/308134/.
24. Roberts, "An Interview with Paul Romer."
25. Steven Landsberg, personal communication at FreedomFest, Las Vegas, 2010.
26. George Dyson, *Turing's Cathedral* (New York: Random House, 2012), 252.
27. Dyson, *Turing's Cathedral*, 252.

CHAPTER ELEVEN: MIND OVER MATTER

This was written as a final chapter, but it seemed the capstone of the "theory" section. It attempts to lift information theory above the level of mere economics. I am mindful, however, of the great Ludwig von Mises' observation that economics "did more to transform human thinking than any other scientific theory before or since." He explained: "With good men and strong governments everything was considered feasible … [but with the advent of economic science] now it was learned that in the social realm too there is something operative which power and force are unable to alter and to which they must adjust themselves if they hope to achieve success, in precisely the same way as they must take into account the laws of nature," as found in Israel M. Kirzner, *Ludwig von Mises* (Wilmington, DE: ISI Books, 2001). I would add the laws of information. All these are paths by which knowledge imposes its laws on the wills and powers of rulers.

1.   James Watson and Andrew Berry, *DNA: The Secret of Life* (New York: Random House, 2009). Introduction: "Our discovery put an end to a debate as old as the human species: Does life have some magical mystical essence, or is it, like any chemical reaction carried out in a science class, the product of normal physical and chemical processes? ... The double helix answered that question with a definitive No.... The double helix is an elegant structure but its message is downright prosaic: life is simply a matter of chemistry." This is the favorite enabling false dichotomy of materialism, ignoring the possibility that life is a product of information and governed not chiefly by chemistry but by information theory.

2.   Matt Ridley, *Francis Crick: Discoverer of the Genetic Code* (New York: HarperCollins, 2012), 107. "Crick was trying to kill a belief that had so far refused to die: the belief that the relationship between DNA and proteins was reciprocal, that DNA determined protein sequences but proteins also determined DNA sequences, and that 'genes' were therefore a combination of both. This was true in a biochemical sense, but it was entirely false in the sense of information. The information required to assemble a protein sequence lay in a DNA sequence; the information required to assemble a DNA sequence also lay in a DNA sequence.... Despite many attempts to topple it, the central dogma remains true: base sequences in DNA determine amino acid sequences in protein, but not vice versa."

3.   Hubert P. Yockey, *Information Theory, Evolution, and the Origin of Life* (Cambridge, UK: Cambridge University Press, 2005), 152–57.

4.   Norbert Wiener, *Cybernetics: or the Control and Communication in the Animal and the Machine*, second edition (Cambridge: MIT Press,1965), 132: "The mechanical brain does not secrete thought 'as the liver does bile,' as the earlier materialists claimed, nor does

it put it out in the form of energy, as the muscle puts out its activity. Information is information, not matter or energy. No materialism which does not admit this can survive at the present day."

5. Albert Einstein, "Maxwell's Influence on the Development of the Conception of Physical Reality," in *James Clark Maxwell, A Commemoration Volume* (Cambridge, UK: Cambridge University Press, 1931).

6. Tom Wolfe, *Back to Blood* (New York: Little, Brown and Company, 2012), 324–325.

7. Jerry Fodor and Massimo Piatelli-Palmarini, *What Darwin Got Wrong* (New York: Farrar, Straus and Giroux, 2010).

CHAPTER TWELVE: THE SCANDAL OF MONEY

This chapter broaches the controversial view—to which I have only lately arrived through my contemplation of information theory—that financial crises are intrinsic to capitalism and key to its success. It is crucial to remedy them without vitiating capitalism.

1. Gary B. Gorton, *Misunderstanding Financial Crises* (New York: Oxford University Press, 2012).

2. Charles Kindleberger, *Manias, Panics, and Crashes*, sixth revised edition (New York: Palgrave McMillian, 2011), with a foreword by Robert Solow. See also Carmen M. Reinhart and Kenneth S. Rogoff, "The aftermath of Financial Crisis" (Working Paper no. 14656, National Bureau of Economic Research, 2009).

3. George Gilder, *Wealth and Poverty, A New Edition for the Twenty-first Century* (Washington, D.C.: Regnery, 2012), 150.

4. Arnold Kling and Nick Schulz, *Invisible Wealth* (New York: Encounter Books, 2011), 225. Superb short analysis of financial crisis.

5. Ronald I. McKinnon, *Money and Capital in Economic Development* (Washington, D.C.: Brookings Institution, 1973).

6. Steve Forbes and Elizabeth Ames, *How Capitalism Will Save Us* (New York: Crown Business, 2009), 242. This book is a lucid and compelling exposition of economic truth. See also, Lewis Lehrman,

The True Gold Standard (New York: The Lehrman Institute, 2011), which makes the case that the U.S. economy is weakened by the unmoored dollar standard.

CHAPTER THIRTEEN: THE FECKLESSNESS OF EFFICIENCY

This chapter revives my theme from *The Spirit of Enterprise* that capitalism cannot be defended without grasping and vindicating the role of capitalists. Vital to capitalist profits are falsifiability: the possibility of bankruptcy. The raptorial revels of bankers with government moneys and guarantees are indefensible by any valid theory of capitalism.

1. Andrew Redleaf and Richard Vigilante, *Panic: The Betrayal of Capitalism by Wall Street and Washington* (Minneapolis, MN: Richard Vigilante Books, 2010).
2. Peter Drucker, *The Effective Executive: The Definitive Guide to Getting the Right Things Done,* revised edition (New York: HarperBusiness, 2006).
3. Andrew Redleaf and Richard Vigilante, *The WhiteBox Investment Advisor* (December 2006).
4. Richard Posner, *The Crisis of Capitalist Democracy* (Cambridge: Harvard University Press, 2010); see also *A Failure of Capitalism* (Cambridge: Harvard University Press, 2009), 84: "big banks increasingly became originators and sellers of debt securities rather than conventional owners of debt until maturity, [and] diversification and value-at-risk models replaced relationship-specific information, as means of controlling risk."
5. Posner, "The Spectre of the Great Depression" in *A Failure of Capitalism* (Cambridge: Harvard University Press, 2009).
6. Posner, "The Spectre of the Great Depression."
7. Michael Lewis, *The Big Short: Inside the Doomsday Machine* (New York: W. W. Norton, 2010).
8. Jay Richards, *Infiltration* (Seattle: Discovery Institute, 2013).

CHAPTER FOURTEEN: REGNORANCE

Jameson Campaign wrote me that any useful book on today's crisis of capitalism must contain a chapter with a theory of regulation. Here it is.

1. Frank Partnoy and Jesse Eisinger, "What's Inside America's Banks?," *Atlantic* (January–February, 2013): 60–71.

2. Friedrich Hayek, *The Road to Serfdom* (Chicago: University of Chicago Press, 2007): 113.

3. John Allison, *The Financial Crisis and the Free Market Cure* (New York: McGraw-Hill, 2012): "The media and other statists have created a myth that the financial crisis was caused by banking deregulation and greed on Wall Street. However, banks were not deregulated. In fact, three major new regulations were passed during the Bush Administration: The Privacy Act, The Patriot Act, and Sarbanes-Oxley. Banks were misregulated, not deregulated."

4. Vern McKinley, *Financing Failure: A Century of Bailouts* (Oakland, CA: The Independent Institute, 2011).

5. Allison, "The Fed's Fatal Conceit," *Cato Journal* 32 no. 2 (Spring–Summer 2012): 265–278.

6. Allison, "The Fed's Fatal Conceit."

7. Allison, "The Fed's Fatal Conceit."

8. Charles Gave, "GDP As a Concept: Misleading if not Outright Criminal," *GaveKal Research* (December 7, 2012).

9. McKinley, *Financing Failure*: 259. Vincent Reinhardt was former director of Monetary Affairs at the Fed.

10. McKinley, *Financing Failure*: 263.

11. McKinley, *Financing Failure*.

12. Neal Freeman, personal communication.

13. Alexander Tabarrok, "The Way Forward" in *Launching the Innovation Renaissance: a New Path to Bring Smart Ideas to Market Fast*, Kindle edition (New York, TED Conferences, 2011). "In the late 20th century we unwisely expanded our patent system. We thought that more monopoly would bring more innovation; it didn't. We need to prune our patent system in order to make room for more growth. We need to make more use of patent buyouts,

prizes, advance market commitments and other innovations in innovation policy."

14. Alan Reynolds, *Income and Wealth* (Greenwood, 2006).
15. Ronald Baker, *Economy of Mind* (New York: Anacom, 2006).
16. Kevin D. Williamson, "The Wonders of Frack," *National Review*, February 20, 2012, 26–31.
17. Williamson, "The Wonders of Frack."
18. Rich Trzupek, *Regulators Gone Wild: How the EPA is Ruining American Industry* (New York: Encounter Books, 2011).
19. Dr. Jane Orient, "EPA v. Human Health," *Civil Defense Perspectives* 29, no. 1 (November 2012): "The EPA is claiming authority to regulate virtually anything it chooses based on the linear no-threshold theory ... [which contends] there is no safe concentration of particulate matter less than 2.5 microns in diameter (PM2.5), as is found in dust storms or diesel exhaust. So far, the EPA has not demanded the use of an N-95 mask when using a vacuum cleaner or duster.... This is a striking inconsistency since indoor PM2.5 levels are much higher than outdoors ... and EPA Administrator Lisa Jackson told Bill Maher that in many areas of the country, 'the best advice is don't go outside. Don't breathe the air. It might kill you."
20. Alexis de Tocqueville, *Democracy in America* (1835 and 1840; reprint, New York: Penguin, 2006): xxxvii.

CHAPTER FIFTEEN: CALIFORNIA DEBAUCH

Here is another chapter on regulation and its effects on the economy. Reading an issue of *Forbes* in 2012 listing "the 100 most important venture capitalists" and their holdings, I noticed that nearly every one gained the bulk of his returns from the Facebook IPO and kindred social networking ventures such as LinkedIn, Groupon, and Zynga. In my view, these companies introduced no formidable new technologies and did not even need venture capital and its special expertise. More valuable than the entire Silicon Valley venture capitalist entente, who were battling for government mandates and subsidies, were the venture capitalists of Washington state, led by Madrona and my friends Tom Alberg and Matt

McIlwain. They successfully fought to defeat a proposed state income tax, perversely supported by Bill Gates and his father. Madrona also is financing better, more promising companies, such as Isilon and Impinj.

Ironically the only winner in Kleiner Perkins's so-called "cleantech" ("green") portfolio is Bryan Mistele's Inrix, which is a Seattle-based "big data" company addressing traffic congestion not through physical changes (fast trains and costly sensors implanted in highways) but through information tools. Inrix's exhaustive real-time data collected from cars, trucks, and GPS-equipped smartphones shows no impact on traffic from mass transit deployments. See Viktor Mayer-Schonberger and Kenneth Cukier, *Big Data* (Boston: Houghton Mifflin Harcourt, 2013). The fact is that the entire onrush of technological progress, especially new cleaner fossil fuels from fracking, advances so-called green goals.

1. "Wealthy Feel Pinch of Trio of Tax Hikes," *San Diego Union-Tribune*, January 6, 2013.
2. Robert Hefner III, *The Grand Energy Transition: The Rise of Energy Gases, Sustainable Life and Growth, and the Next Great Economic Expansion* (New York: John Wiley & Sons, 2011).
3. Thomas Friedman, *Hot, Flat, and Crowded: Why We Need a Green Energy Revolution—and How It Can Renew America* (New York: Farrar, Straus and Giroux, 2008). Max Schulz offers a chilling account of what California's green policies are accomplishing in "California's Potemkin Environmentalism," *City Journal*, Spring 2008. On the cost of AB32's CO_2 targets: "Consider that California could take every one of its 14 million passenger cars off the road, and still be less than halfway toward its goal.... Shutting down 100 state-of-the-art natural-gas-fired power plants still wouldn't get us there. Closing the entire cement industry ... wouldn't finish the job."
4. Eugene Fitzgerald, *Inside Real Innovation* (Cambridge: MIT Press, 2010).
5. William Tucker, "The Energy Crisis is Over! How We Beat OPEC," *Harper's*, November 1981.

6.   Matt Ridley, *The Rational Optimist* (New York: HarperCollins, 2010): 241.

7.   Peter Huber, *Hard Green: Saving the Environment from the Environmentalists* (New York: Basic Books, 1999). This was the first critique of environmentalism to demonstrate that the "green" movement is the world's chief threat to the environment because it squanders precious land to obviate use of abundant subterranean resources extractable with little or no permanent damage to the environment.

8.   Ridley, *The Rational Optimist*, 243. In early 2013, Ridley achieved a breakthrough by calculating that the rate of plants' absorption of CO_2 exceeded the rate of incremental heat capture by atmospheric CO_2. Thus consumption of fossil fuels has the net effect of making the planet greener. Particularly benefited is the Amazon rain forest, which has shown anomalous growth in recent years in proportion to the growth of CO_2 in the atmosphere. Read more online at http://www.aei-ideas.org/2013/03/matt-ridley-burning-fossil-fuels-is-greening-the-planet/.

CHAPTER SIXTEEN: DOING BANKING RIGHT

This chapter was prompted by my discovery that my neighbor Robert Wilmers had steered his bank, M&T, through the financial crisis with nary a down quarter, in the face of massive subsidies for his failing rivals. Then I found that John Allison had achieved a similar success. What was their secret? They favored knowledge over power.

1.   William M. Isaac and Phillip C. Meyer with foreword by Paul Volcker, *Senseless Panic: How Washington Failed America* (New York: John Wiley & Sons, 2012): 104.

2.   Redleaf and Vigilante, *Panic: The Betrayal of Capitalism by Wall Street and Washington* (Minneapolis, MN: Richard Vigilante Books, 2010): 179.

3.   Nassim Nicholas Taleb and Mark Spitznagel, in a blog post at CNN Public Square from October 2012, estimate that $2.2 trillion was paid to bankers, chiefly in bonuses, in the U.S. alone between

June 2000 and June 2007 and projected the total to rise to close to $5 trillion over the course of the decade. "Bankers used leverage to increase profitability and exploited the backstop of public guarantees. The profits largely flow to the employees [i.e. the bankers], while the losses are defrayed by the taxpayers and shareholders and even retirees (through artificially low interest rates). The Fed also provided $1.2 trillion in loans to banks (mostly secret at the time)."

4. Don Luskin and Andrew Greta, *I Am John Galt* (Hoboken, NJ: John Wiley & Sons, 2011).

5. John Allison, *The Financial Crisis and the Free Market Cure* (New York: McGraw-Hill, 2012).

6. Lewis E. Lehrman, "Money for Nothing," review of *The Financial Crisis and the Free Market Cure* by John Allison, *Weekly Standard,* January 14, 2013, http://www.weeklystandard.com/articles/money-nothing_693747.html.

7. Allison, *The Fed's Fatal Conceit* (Washington, D.C.: Cato Institute, Spring/Summer 2012): 265.

CHAPTER SEVENTEEN: THE ONE PERCENT

I have been writing this chapter for nearly thirty years. The message is perennial. The Kessler quote comes from the *Wall Street Journal.*

1. Jeffrey Sachs, *The End of Poverty: Economic Possibilities for Our Time* (New York: Penguin Press, 2005).

2. Jeremy Rifkin, *The European Dream: How Europe's Vision of the Future Is Quietly Eclipsing the American Dream* (New York: Tarcher, 2005).

3. Richard Wilkinson and Kate Pickett, *The Spirit Level: Why Equality is Better for Everyone* (London: Bloomsbury Press, 2011). *The Spirit Level* was effectively refuted before publication by Arthur C. Brooks of the American Enterprise Institute, who provides similar graphs registering the happiness of free people compared to unfree: "Free People Are Happy People," *City Journal,* Spring 2008, http://www.city-journal.org/2008/18_2_happy_people.html.

4. Kevin D. Williamson, "Everybody Gets Rich," *The New Criterion* 30, no. 5 (January 2012): 4–12. Williamson, our new Tom Wolfe, asks the crucial question: "From energy used, to calories consumed, to travel enjoyed, to the size of our houses, to the variety of our diets and distractions, we are rich, rich, rich, besotted with wealth, drowning in affluence, up to our fat little earlobes in the good life. So why do we feel so poor?"

5. Eric Schmidt wrote me in an email 1993 that "when the network runs as fast as the computer backplane, the computer will hollow out and distribute itself around the network, and profits in the industry will migrate toward the providers of 'sort' and 'search' capabilities." I called the insight "Schmidt's Law." Sure enough, less than a decade later, network speeds passed the barrier and he became CEO of Google. Chris Cooper of Seldon Technologies in Windsor, Vermont, invented a tunable "nanomesh" based on carbon nanotubes that can produce fresh potable water from a septic pool without power or chemicals (I am on the board of his company). Jules Urbach of Otoy corporation in Los Angeles invented and learned to program petaflop computers based on clustered graphics processors that could compensate for speed of light latency and render real-time 3-D graphics from the "Cloud" for Autodesk and Hollywood.

CHAPTER EIGHTEEN: THE BLACK SWANS OF INVESTMENT

Taleb rather irritated me with *Fooled by Randomness* and *The Black Swan*, which seemed to me to flout a key tenet of information theory: it is impossible in principle to distinguish a series of random data points from a series of creative surprises. That Wall Street and entrepreneurial returns seem random is irrelevant to the question of whether they were earned by creative contributions or through mere luck. *Antifragile*, however, seems to me to be an altogether different and far more interesting story.

1. Nassim Nicholas Taleb, *Fooled by Randomness* (New York: Random House, 2001).

2. Taleb, *The Black Swan* (New York: Random House, 2007).

3. Ken Fisher, *The Only Three Questions That Count*, revised edition (2007; reprint, New York: John Wiley & Sons, 2012). He also wrote, with Lara Hoffmans, *Debunkery: Learn It, Do It, and Profit From It—Seeing through Wall Street's Money Killing Myths* (New York: John Wiley & Sons, 2011), which followed *Three Questions* in debunking the impact of Black Swans, national debt, the consumer as king, and even high capital gains taxes (which is tax *selling*, not holding shares).

4. Gregory Chaitin, *Proving Darwin: Making Biology Mathematical* (New York: Vintage, 2013).

5. Shannon, "*Omni* interview" in *Collected Papers*.

6. Taleb, *Antifragile: Things That Gain from Disorder* (New York: Random House, 2012).

7. Clayton Christensen, *The Innovator's Dilemma*, revised edition (1997; reprint, New York: HarperBusiness, 2012).

CHAPTER NINETEEN: THE OUTSIDER TRADING SCANDAL

This is another old favorite theme that I have been touting for decades. Some day the world will come around to seeing that the idea of regulating inside trading is futile and destructive because it distorts the information and suppresses the knowledge that makes markets work. Fraud is another matter that can be managed by existing laws against it.

1. Burton Malkiel, *A Random Walk Down Wall Street,* revised and updated edition (New York: W. W. Norton, 2007).

2. John Mauldin, edit., *Just One Thing* (New York: John Wiley & Sons, 2006).

3. Benoit Mandelbrot and Richard L. Hudson, *The Misbehavior of Markets: A Fractal View of Risk, Ruin, and Reward* (New York: Basic Books, 2006).

4. Thomas Cover and Joy Thomas, *Elements of Information Theory*, second edition (New York: John Wiley & Sons, 2006).

5.   Cover, *Elements of Information Theory.*
6.   Robert Laughlin, *A Different Universe* (New York: Basic Books, 2005).
7.   Fisher, *Debunkery*, 182 and passim.
8.   Richard Vigilante, personal communication based on *Panic: The Betrayal of Capitalism by Wall Street and Washington* (Minneapolis, MN: Richard Vigilante Books, 2010).

CHAPTER TWENTY: THE EXPLOSIVE ELASTICITIES OF FREEDOM

One of my inspirations for this chapter was reading John Train's *The Unruly Monkey: Reflections on Life, Love, and Money* (Maria Teresa Train MTT Scala, 2012), a scintillating collection of sage observations from a lifetime as a writer and investor: "*Train's First Law*: Price Controls Increase Prices … by inhibiting production.… *Train's Second Law*: The more the government does something the less there is of it. The government costs about twice as much and takes three times as long as the private sector to do any given thing.… At the limit, under Communism, the government does almost everything, and people receive little indeed. 'We pretend to work and they pretend to pay us,' said Soviet workers. *Train's further Law*: Make-work Programs Reduce Employment [as] their cost consumes the productive investment that creates real jobs."

Because of Train's Laws, retrenchment of government spending and regulation releases explosive energies of growth.

1.   Peter Drucker, *The Effective Executive* (New York: Harper & Row, 1966): 104. "Executives … are forever bailing out the past.… Yesterday's actions and decisions … inevitably become today's problems, crises, and stupidities."
2.   David Stockman, "Paul Ryan's Fairy Tale Budget Plan," *New York Times*, August 13, 2012, http://www.nytimes.com/2012/08/14/opinion/paul-ryans-fairy-tale-budget-plan.html?_r=0.
3.   Stockman, "David Stockman Rips Romney on Jobs and the Bain Drain," *Newsweek*, October 22, 2012, 25–31. Found in *The Great*

Deformation: How Crony Capitalism Corrupts Free Markets and Democracy (New York: Perseus Books, 2013).

4. Alan Reynolds, "David Stockman vs. Bain Capital," National Review Online, October 22, 2012, http://www.nationalreview.com/articles/331128/david-stockman-vs-bain-capital-alan-reynolds.

5. Nick Gillespie, "The Triumph of Politics over Economics: David Stockman on TARP, the Fed, Ronald Reagan, and Ron Paul," *Reason*, April 1, 2011, 22ff.

6. Mancur Olson, *The Rise and Decline of Nations* (New Haven: Yale University Press, 1982).

7. Drucker, *The Effective Executive*, 105. "Government programs and activities age just as fast as the programs and activities of other institutions. Yet ... they are welded into the structure through civil service rules and immediately become vested interests.... At a guess, at least half the bureaus and agencies ... either regulate what no longer needs regulation ... or they are directed, as is most of the farm program, toward investment in politicians' egos and toward efforts that should have had results but never achieved them. There is a serious need for a new principle of effective administration under which every act, every agency, and every program of government is conceived as temporary and as expiring automatically after a fixed number of years—maybe ten—unless specifically prolonged by new legislation following careful outside study...."

8. Thomas Friedman and Michael Mandelbaum, *That Used to be Us: How America Fell Behind in the World It Invented and How We Can Come Back* (New York: Farrar, Straus and Giroux, 2011).

9. Thomas Macaulay, *The History of England* (London: Longmans, Green, Reader, and Dyer, 1871). This is a two-volume edition inscribed by my great-grandfather in 1878. There are newer, more accessible versions available in digital form.

10. Macaulay, "Chapter 19" in *The History of England*, 398–402. This passage contains some of the greatest writing in the history of economic literature.

11. Arnold Kling, *Unchecked and Unbalanced: How the Discrepancy Between Knowledge and Power Caused the Financial Crisis and*

Threatens Democracy (Lanham, MD: Rowman & Littlefield, 2010). A shorter version is Jason Taylor and Richard Vedder, "Stimulus by Spending Cuts, 1946," *Cato Institute Report* (August 2010): "Civilian employment grew, on net, by over 4 million between 1945 and 1947 when so many pundits were predicting economic Armageddon. Household consumption, business investment, and net exports all boomed as government spending receded. The postwar era provides a classic illustration of how government spending 'crowds out' private sector spending and how the economy can thrive when the government's shadow is dramatically reduced."

12. Mark Skousen, "Measures of Economic Activity, Income and Wealth" in *Economic Logic*, revised third edition (Washington, D.C.: Capital Press, 2010), 337–354.

13. Robert Higgs, "Wartime Prosperity? A Reassessment of the U.S. Economy in the 1940s," The Independent Institute, March 1, 1992, http://www.independent.org/publications/article.asp?id=138. Also, Higgs, "Regime Uncertainty: Why the Great Depression Lasted So Long and Why Prosperity Resumed after the War," *Independent Review*, no. 4 (1997): 561–590.

14. Maurice McTigue, "Rolling Back Government, Lessons from New Zealand," *Imprimis*, Hillsdale College (February 11, 2004).

15. George Gilder, *The Israel Test* (New York: Encounter Books, 2012), tells the story.

16. Canada's experience also reaps the benefit of a strong Canadian dollar, while U.S. weakness feeds on U.S. dollar decline. See David Malpass, "Sound Money, Sound Policy" in *The 4% Solution: Unleashing the Economic Growth America Needs,* edit. Brendan Miniter (New York: Crown Business, 2012), 94–107. "Brazil's economic boom dates from 2003, when President Luiz Inacio Lula da Silva decided to stop the real's collapse.... Russia's frenzied economic collapse ended almost overnight when ... Putin ... stabilized the ruble. [Similarly in China, economic growth surge] started in mid-1993 after then-vice premier Zhu Rpongji ... moved his office to the central bank to stop the communist-era black

market weakness in the renmimbi.... The current dollar policy creates a circular growth crisis in which dollar weakness discourages investment in dollars.... The investment killer is that assets in the United States keep losing value in foreign-currency terms, so potential U.S. investments start with a big rate-of-return disadvantage relative to other countries with better currency policies."

17. Fraser Nelson, "Sweden's Secret Recipe," *Spectator*, April 14, 2012.

CHAPTER TWENTY-ONE: FLATTENING TAXES

My original title for this chapter was "Is Economic Growth Caused by Taxes? Or by Global Warming?" It was provoked by the timely Congressional Research Service "study" released in mid-October 2012. Sorry, I know, it's now "climate change."

1. Thomas Hungerford, "Taxes and the Economy: An Economoc Analysis of the Top Tax Rates Since 1945," Congressional Research Service, December 12, 2012, http://democrats.waysandmeans. house.gov/sites/democrats.waysandmeans.house.gov/files/ Updated%20CRS%20Report%2012%3A13%3A12.pdf.
2. In the second decade of the new millennium, Laffer was still at it, producing a unique and definitive analysis of the deadweight loss inflicted on the economy through the excessive complexity of the tax system. Arthur Laffer, Wayne Winegarden, and John Childs, *The Economic Burden Caused by Tax Code Complexity* (Laffer Associates: Knoxville, TN, April 2011).
3. Keith Marsden, "Taxes and Growth," *Finance and Development* 20 (September 1983), 40–43. Appraising the economic performance of twenty countries grouped in pairs of similar per capita incomes in 1970, the study found that the countries with lower effective tax burdens grew six times faster. Thus they could raise their government spending and revenues three times as fast as the countries with higher tax burdens.
4. Jude Wanniski and David Goldman, with Jay Turner and Evan Kalimtgis, "A Flat Tax Would Produce Explosive U.S. Economic

Growth" in *Supply Side Analytics* (Morristown, NJ: Polyconomics, 1992). Refining the Marsden analysis and extending it through the 1980s, the Polyconomics study found that marginal tax rates explained between 80–90 percent of differential growth rates among the world's twenty-four leading economies. Fastest was Hong Kong, with a top rate of 15.9 percent; slowest was Denmark with an 80.2 percent top rate. Capital gains taxes were shown to display a particularly acute inverse ratio between rates and revenues.

5.   Chris Edwards and Daniel Mitchell, *The Global Flat Tax Revolution* (Washington, D.C.: Cato Institute, 2008).

6.   The flat tax movement was given its major push by Steve Forbes' presidential campaign. See Steve Forbes, *The Flat Tax Revolution: Using a Postcard to Abolish the IRS* (Washington, D.C.: Regnery, 2005).

7.   Kevin Hassett and James K. Glassman, "Spending, Taxes, and Certainty: A Road Map to 4%," in *The 4% Solution: Unleashing the Economic Growth America Needs*, Edited by Brendan Miniter (New York: Crown Business, 2012), 80–93.

CHAPTER TWENTY-TWO: THE TECHNOLOGY EVOLUTION MYTH

This chapter embodies the origin and provocation for *Knowledge and Power*. Having innocently reveled in his previous work, *Out of Control*, I received Kevin Kelly's cornucopian new book, *What Technology Wants*, with high expectations, and mounted its rhetorical roller coaster with high enthusiasm. However, my sense of anticipation gradually gave way to a sinking feeling that there was something wrong. The book turned out to be part of a movement among technology writers to reduce technology and its creators to epiphenomena of a physical and biological "technium." President Obama was just boarding this bandwagon when he threw down his gauntlet to entrepreneurs: "You did not build that." In Kelly's pages individual inventors and entrepreneurs virtually disappear into an evolutionary "river of mutations" that "can be traced back to the life of an atom ... created in the fires of the big bang." The book ended with a kind of empty eloquence: "The expanding technium—its cosmic trajectories, its ceaseless reinvention, its inevitabilities, its self-generation—is an open-ended beginning, an infinite game calling us to play." His prose was inspiring, but his blur of information theory into a botch of entropy and "exotropy," order and self-organization, gave me the push to write this book.

1.   Kevin Kelly, *What Technology Wants* (New York: Viking, 2010).
2.   Steven Johnson, *Where Good Ideas Come From* (New York: Penguin, 2010).
3.   George B. Dyson, *Darwin among the Machines: The Evolution of Global Intelligence* (Reading, MA: Addison-Wesley, 1997).
4.   George B. Dyson, *Turing's Cathedral* (New York: Pantheon Books, 2012). A dense intractable book that rewards a close reading. It tells the story of the evolution of computing from individual machines laboriously wrought from tubes and relays at the Institute for Advanced Studies to virtual machines in the "cloud" with many trillionfold gains in efficiency and accessibility in prospect.
5.   Johnson, "Conclusion: The Fourth Quadrant" in *Where Good Ideas Come From*, unindexed Kindle edition.
6.   James Besson, Jennifer Ford, and Michael J. Meurer, "The Private and Social Costs of Patent Trolls," (Working Paper 11–45, Boston University School of Law, September 19, 2011). This is a good first take of the argument, though it does not distinguish between the costs of the issuance of trivial patents and the damage from litigation.
7.   Kelly, "Seeking Conviviality" in What Technology Wants, 239–265.
8.   Kelly, *What Technology Wants*, 16–17.
9.   John Searle, *The Mystery of Consciousness* (New York: The New York Review of Books, 1990), 109 and passim.
10. Kelly, "Part 4: Directions" in *What Technology* , 325–333 and passim. But "unconscious free will" is as meaningless as an unconscious mind or a bottom-up god.

11. Kelly, *What Technology Wants*, 57.
12. Dyson, *Darwin among the Machines*, 10. "For those who fear mechanistic explanations of the human mind, our ignorance of how local interactions produce emergent behavior offers a reassuring fog in which to hide the soul."
13. Kelly, *What Technology Wants*, 333.
14. Albert O. Hirschman, "The Principle of the Hiding Hand," *Public Interest*, no. 6 (Winter 1967).
15. Peter Drucker, *The Effective Executive: The Definitive Guide to Getting the Right Things Done*, revised edition (New York: HarperBusiness, 2006).
16. Max Delbruck, *Mind From Matter? An Essay on Evolutionary Epistemology* (Blackwell, 2005), 269.
17. Robert Laughlin, *A Different Universe* (New York: Basic Books, 2005), 213 and passim.

CHAPTER TWENTY-THREE: ISRAEL: INFONATION

The Israel Test (2009; reprint, Encounter Books, 2012) elaborates on these themes.

Tom Bethell, in his scintillating biography of *Eric Hoffer: The Longshoreman Philosopher* (Stanford: Hoover Institution Press, 2012), 277–278, publishes Hoffer's lapidary words on Israel: "[A]t this moment [1968 and even more today], Israel is our only reliable and unconditional ally. We can rely more on Israel than Israel can rely on us.... I have a premonition that will not leave me; as it goes with Israel so will it go with all of us."

1. Kevin Fitchard, "It's Alive! AT&T's Networks Become Self-Aware," GigaOM, February 12, 2012, http://gigaom.com/2012/02/24/its-alive-atts-networks-become-self-aware/.
2. Arthur Herman, "How Israel's Defense Industry Can Help Save America," *Commentary*, December 2011, 14–19.

CHAPTER TWENTY-FOUR: THE KNOWLEDGE HORIZON

In recent years, I have found myself debating the superb venturer and libertarian thinker Peter Thiel (See for example "Peter Thiel and George Gilder debate on 'The Prospects for Technology and Economic Growth,'" Youtube video, 1:16:27, posted by Intercollegiate Studies Institute "EducatingForLiberty," March 14, 2012, http://www.youtube.com/watch?v=XRrLyckg8Nc.) I have learned much from his sharp insights, which have forced me toward hard analysis of the foibles of contemporary science and technology and the promise of the new information tools.

1.   Tyler Cowen, "The Low-Hanging Fruit We Ate" in *The Great Stagnation: How America Ate All the Low-Hanging Fruit of Modern History, Got Sick, and Will (Eventually) Feel Better*, Kindle edition (New York: Dutton, 2011). The quotes are in Chapter One; Kindle edition lacks page numbers and has yet to be indexed. Peter Thiel's more erudite expressions of innovation angst run from "The Optimistic Thought Experiment," *Policy Review*, no. 147 (January 29, 2008) to "The End of the Future," *National Review*, October 3, 2011.

2.   Andy Kessler has written extensively and uproariously on technology and finance through seven savvy and insightful books and indispensable articles in the *Wall Street Journal* and *Forbes*. The books include *Wall Street Meat* (2001), *Running Money* (2003), *The End of Medicine* (2007), *How We Got Here* (2008), *Eat People (2011)*, *How to Kick Ass on Wall Street* (2013), and the riotously savvy and only slightly fictional futurism of *Furby* (2012). Since some of his work is self-published (they get picked up later by HarperCollins et al), find them all, and much more, on his website at www.andykessler.com. This quote comes from "The Rise of Consumption Equality," *Wall Street Journal*, January 3, 2012.

3.   Cowen, The Great Stagnation.

4.   Thiel, ibid.

5.   Charles Murray, *Human Accomplishment: The Pursuit of Excellence in the Arts and Sciences, 800 BC to 1950* (New York: Harper, 2003). A great book.

6. Ludwig von Mises, "A Draft of Guidelines for the reconstruction of Austria" in *Selected Writings of Ludwig von Mises: The Political Economy of International Reform and Reconstruction*, edited and with an introduction by Richard M. Ebeling (Indianapolis: Liberty Fund, 2000). Cowen, *The Great Stagnation*. Also addressing this issue is Terence Kealey, *Sex, Science & Profits* (London: William Heinemann, 2008). A path-breaking book with a botched title. Sex turns out to have nothing to do with it, beyond the obvious, but profits turn out to be indispensable to scientific advance.

7. George Gilder, *Wealth and Poverty* (Washington, D.C.: Regnery, 2012), 6.

8. Neal Stephenson, "Innovation Starvation" in *Some Remarks* (New York: William Morrow, 2012), 313. Stephenson is my favorite contemporary novelist and he has written more profoundly about science and technology than any other writer of literature.

9. Thiel, "The End of the Future."

10. Richard Feynman, "There is Plenty of Room at the Bottom" (lecture given to the American Physical Society, December 1959), *Engineering and Science* (Caltech, 1960).

11. For just a few accessible samples of the infinite parallel universes genre, which keeps on giving: Steven Weinberg, *Dreams of a Final Theory* (New York: Vintage, 1994); Michio Kaku, *Hyperspace: A Scientific Odyssey Through Parallel Universes, Time Warps and the Tenth Dimension* (New York: Anchor, 1995); David Deutch, *The Fabric of Reality: The Science of Parallel Universes—and Its Implications* (New York: Penguin Books, 1998); Michio Kaku, *The Physics of the Future: How Science Will Shape Human Destiny and our Daily Lives by the Year 2100* (New York Doubleday, 2011). (In case you are interested, Michio thinks it involves multiple parallel universes galore.)

12. Seth Lloyd, *Programming the Universe* (New York: Knopf, 2005).

13. Gregory Chaitin, *Conversations with a Mathematician: Math, Art, Science, and the Limits of Reason,* (New York: Springer, 2002). The relative lack of information content in these disciplines is discussed by Hubert P. Yockey, *Information Theory, Evolution, and*

The Origin of Life, (New York: Cambridge University Press, 2005), 2, 169.

14. Christopher A. Fuchs, *Coming of Age with Quantum Information* (Cambridge, UK: Cambridge University Press, 2011), 133.
15. Fuchs, *Coming of Age with Quantum Information*, 133
16. Fuchs, *Coming of Age with Quantum Information*, 531. For a history of quantum information, see Louisa Gilder, *The Age of Entanglement* (New York: Knopf, 2008).

Note on cut text: Carver Mead, Collective Electrodynamics (Cambridge: MIT Press, 2005). An elegant and powerful vision of a different path into the mysteries of matter, offered by the Gordon A. Moore Professor of Science and Engineering at Caltech after fifty years exploring matter from the inside in tunnel diodes, microchips, and biomorphic electronic circuitry. His story is told in my Microcosm (Simon and Schuster, 1999) and The Silicon Eye (New York: W. W. Norton, 2006).

CHAPTER TWENTY-FIVE: THE POWER OF GIVING

This is an old favorite theme from *Wealth and Poverty*, revised and updated to focus on the *entropy* of giving—how gifts are most gratifying when they reveal surprising and unexpected sympathies.

Thus gifts resemble profits, which are an index of altruism and surprise.

1. Harvey Liebenstein, *Beyond Economic Man* (Cambridge: Harvard University Press, 1976).
2. Burton H. Klein, *Dynamic Economics* (Cambridge: Harvard University Press, 1977).

Index